SB2C Helldiver

AF682070

Written by David Doyle

In Action®

Color Art by Don Greer

Line Illustrations by Todd Sturgell

Squadron/Signal® Publications

(Front Cover) The last dive-bomber to enter operational service with the U.S. Navy, the Curtiss SB2C Helldiver compiled an impressive combat record in the Pacific Theater. (Stan Piet collection)

(Back Cover) Curtiss SB2C number 110 of Bombing Squadron 3 (VB-3) participates in an attack on an IJN warship in early 1945. The plane wears the diagonal white tail marking and white stripes on the wings that were the identification signs of that unit.

About the In Action® Series

In Action® books, despite the title of the genre, are books that trace the development of a single type of aircraft, armored vehicle, or ship from prototype to the final production variant. Experimental or "one-off" variants can also be included. Our first *In Action*® book was printed in 1971.

ISBN 978-0-89747-766-6

Proudly printed in the U.S.A.
Copyright 2015 Squadron/Signal Publications
1115 Crowley Drive, Carrollton, TX 75006-1312 U.S.A.
All rights reserved. No part of this publication may be reproduced, stored in a retrieval system, or transmitted in any form by means electrical, mechanical, or otherwise, without written permission of the publisher.

Military/Combat Photographs and Snapshots

If you have any photos of aircraft, armor, soldiers, or ships of any nation, particularly wartime snapshots, why not share them with us and help make Squadron/Signal's books all the more interesting and complete in the future? Any photograph sent to us will be copied and returned. Electronic images are preferred. The donor will be fully credited for any photos used. Please send them to:

Squadron/Signal Publications
1115 Crowley Drive
Carrollton, TX 75006-1312 U.S.A.
www.SquadronSignalPublications.com

(Title Page) A Curtiss SB2C-5 Helldiver, the U.S. Navy's last dive bomber, is brought up to the flight deck of CV-47, the USS *Philippine Sea* in early 1948. The aircraft was assigned to squadron VA-9a, the "Tough Kitties."

Acknowledgments

This book would not have been possible without the generous help of my friends, including Tom Kailbourn, Stan Piet, Scott Taylor, the staffs of the National Museum of Naval Aviation, Naval Historical Center, the National Archives and the National Museum of the United States Air Force. The dedicated staff at Squadron Publications were supremely helpful, as they always are, in presenting this material in a clear, concise manner. As always, none of this would have been possible without the support and devotion of my wife Denise.

Unless otherwise noted, all the photos in this volume are from the collection of the National Museum of Naval Aviation.

Introduction

By the late 1930s the U.S. Navy earnestly desired a replacement for its proven and reliable Douglas SBD Dauntless, a design that dated back to 1934. On 29 June 1938 the Navy Bureau of Aeronautics issued a set of specifications for a new dive bomber. The requirements included the capability to carry a 1,000-pound bomb load 1,000 miles; being able to take off from an aircraft carrier without the use of a catapult; being able to land on a carrier or on the ground; being able to accommodate a two-man crew, who were protected by armor; and finally, it was preferable that two of the new aircraft fit simultaneously, with one foot to spare, on an aircraft elevator measuring 41 by 48 feet. The aircraft was to be powered by the then-new Wright R-2600 Cyclone engine.

In August 1938, invitations for bid were solicited, and among the six responses was a design submitted by Curtiss-Wright on 6 December as its Model 84. Serious consideration soon focused on the Curtiss offering and on that of Brewster, which had proposed the aircraft that would become the SB2A Buccaneer. Brewster and Curtiss both received contracts for a single aircraft each. Brewster's contract was awarded on 4 April 1939, and Curtiss's on 15 May. The Curtiss aircraft was designated SB2C – indicating that it was the second Curtiss (2C) scout bomber (SB) developed for the Navy.

The first SB2C was unable to fly – it was a wooden mockup constructed at the company's Buffalo plant and inspected by the Navy in late May 1939. As originally designed, and as exhibited in this wooden prototype, the Helldiver was to have had an enclosed turret for the rear gunner. But the installation of the turret was to prove problematic, and despite constructing two additional mockups, no reasonable solution could be found. In February 1940 the Navy agreed to consider a simpler flexible mounting for two .30-caliber machine guns to be operated by the rear-seat man.

The design continued to be refined through wind tunnel tests into early 1940. These tests revealed a number of problems with stability and weight. Among the changes made to correct these issues were moving the engine forward a foot, completely redesigning and enlarging the wing, and incorporating aluminum forgings and magnesium components.

The problems of weight, stability, and strength continued to plague all the early Helldiver models. Added to these issues were quality control concerns and late delivery – so late such that four years after contract not one combat-ready aircraft had been delivered.

The aircraft's problems were so severe that they brought on an investigation by the famed Truman Committee. But by the time the committee issued its scathing report in 1943, most of the Helldiver's teething problems had been worked out, and VB-4 and VB-6 were aboard second *Yorktown* on her shakedown cruise. There were still enough problems, however, that *Yorktown's* Captain, J.J. "Jocko" Clark, recommended the program be cancelled, and *Yorktown* sailed for the combat zone with the veteran Dauntless.

Soon the problems were resolved, and the big bomber entered combat, first flying from the *Bunker Hill.* In short order the Helldiver was on its way to racking up an impressive list of victories in the Pacific. Among these were being instrumental in sinking the two largest battleships ever built, Japan's *Yamato* and *Musashi.*

The SB2C was the second Curtiss aircraft to bear the Helldiver moniker. The earlier Helldiver was a variant of the F8C-3 built in 1928. The fabric-covered 450-horsepower biplane is a far cry from the 2,000+ horsepower monoplane that is the subject of this volume. (National Museum of Naval Aviation)

A plane captain sits in the cockpit of an SBD-3 (15) aboard a carrier in 1942. Squadron number and function letters were generally removed from carrier-based aircraft by this time. Circular holes in the Dauntless' flaps reduced air pressure on the flaps when they served as air brakes during diving attacks. (USN via W.A. Riley)

SB2C Development

XSB2C-1 (1st version)

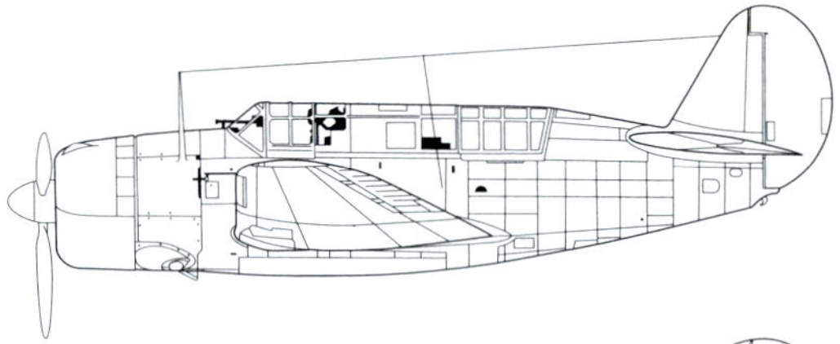

XSB2C-1 (2nd version)

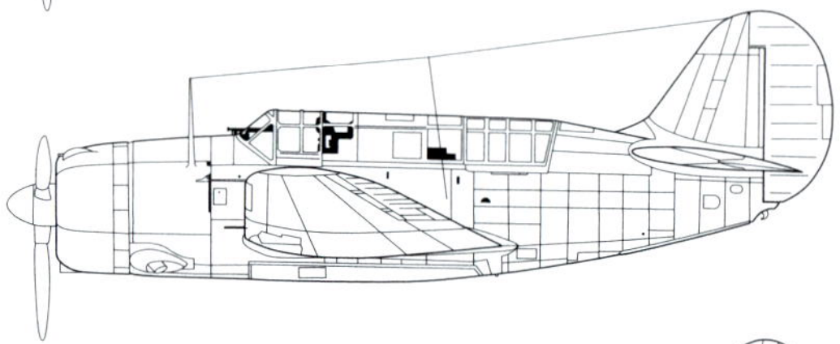

SB2C-1

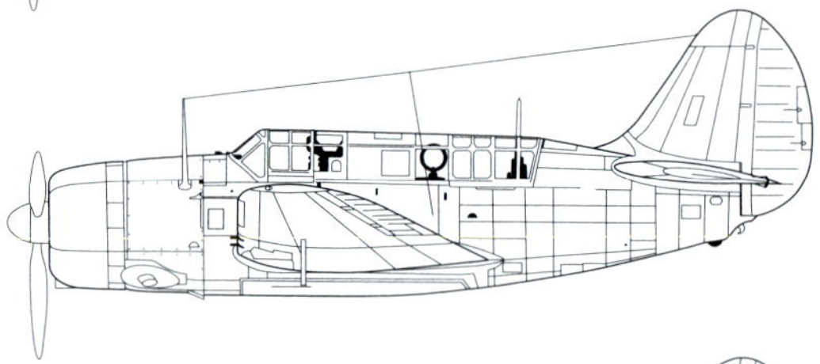

SB2C-1 Mod II

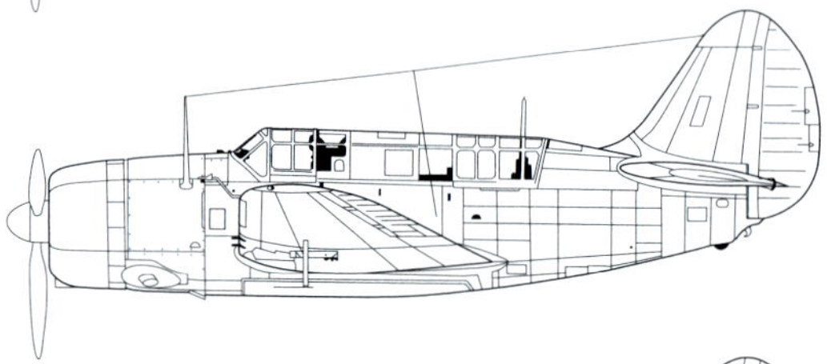

SB2C-1 Mod III

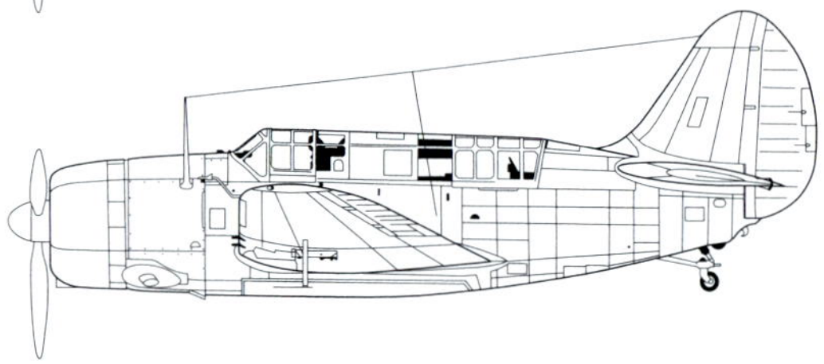

A-25A

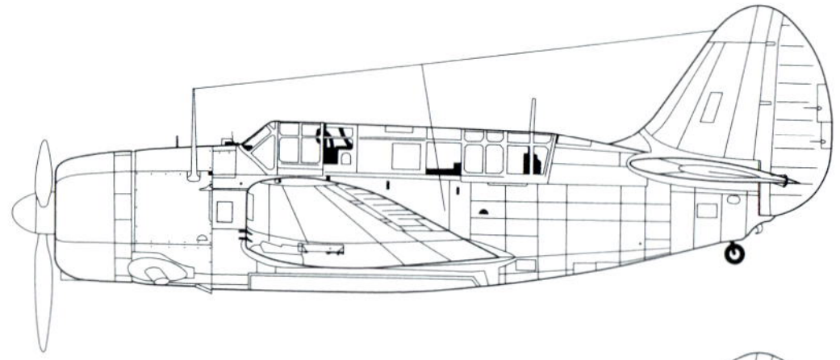

SB2C-1C

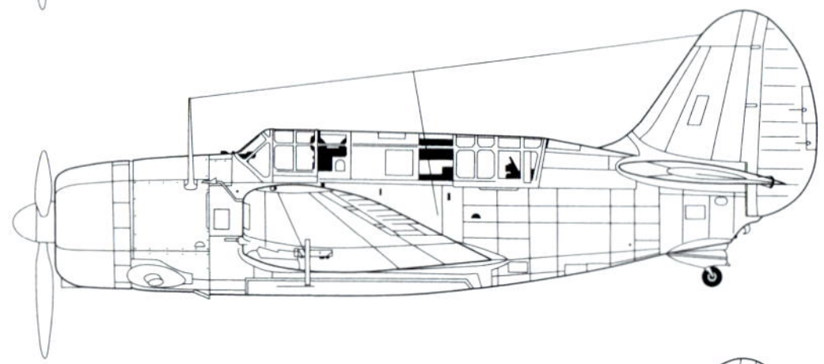

SB2C-3

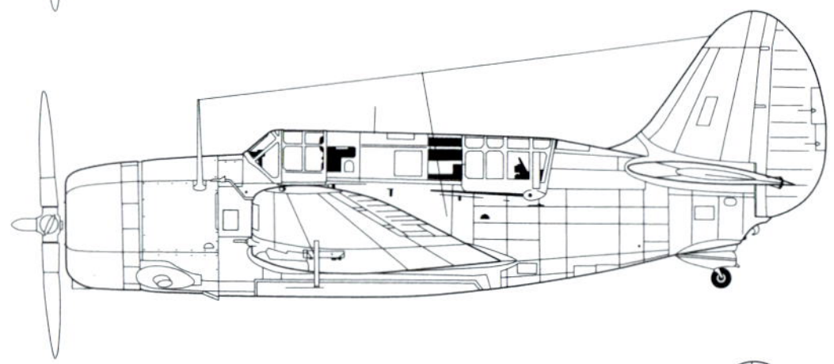

SB2C-4E

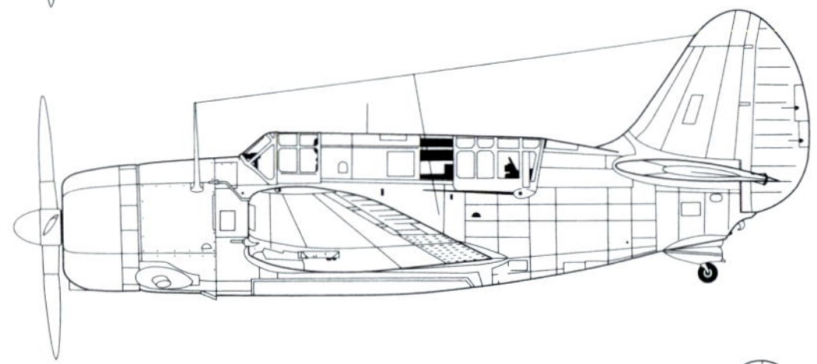

XSB2C-6

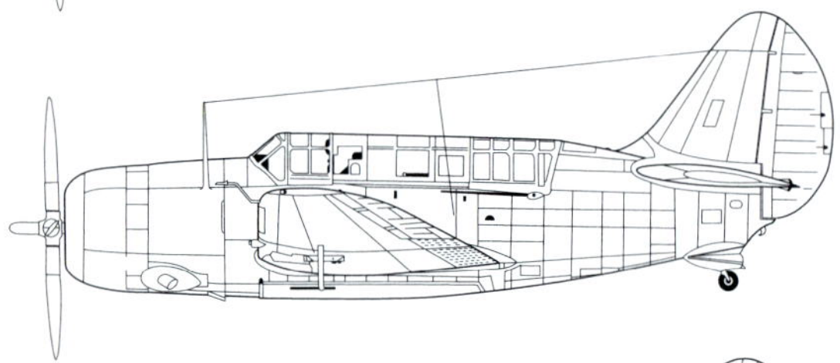

SB2C-5E

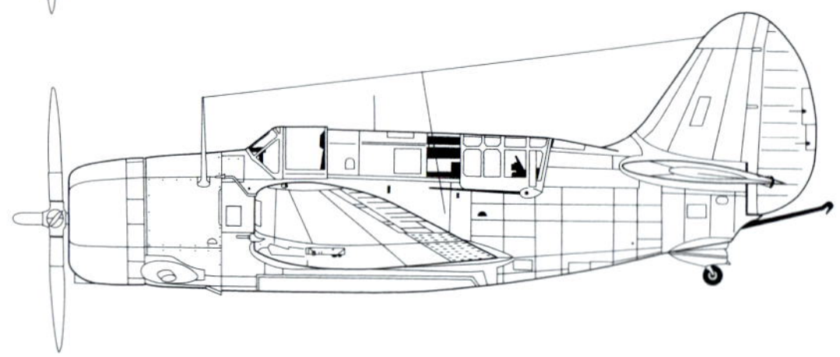

XSB2C-1

In May 1939 the U.S. Navy ordered from Curtiss-Wright Corporation (hereafter referred to as Curtiss) a single prototype of a single-engine dive bomber with a bomb bay. This aircraft, designated the Curtiss XSB2C-1, was assigned U.S. Navy Bureau Number (BuNo) 1758. By the time the prototype was completed on 13 December 1940, the Navy had placed an order with Curtiss for 370 production SB2C-1s. The prototype's first flight was on 18 December 1940. It is shown in an overall aluminum paint scheme, with yellow wing tops.

The XSB2C-1, Bureau Number 1758, was completed and weighed on 2 December 1940, tipping the scales at 7,122 pounds empty. Unfortunately, its initial test flight, on 18 December, revealed that despite the wind tunnel testing not all the problems had been solved. The aircraft lacked proper engine cooling and some of its hatches were weak but the main concern was the plane's lack of stability, augmented by buffeting while diving.

Flight testing came to a halt on 9 February 1941 when there was an in-flight engine failure during the 26th test flight. The pilot managed to land the plane, but the landing was heavy and the aircraft broke in two at the turtle deck. The rebuilding subsequent to this crash incorporated modifications hoping to address many of the previously revealed issues, and the reborn bomber again lifted into the air on 6 May 1941. The improvements thus far had pushed the empty weight of the aircraft to 8,030 pounds. Only a few days later, on 10 May, the left landing gear failed, sending the big bird back into the shop. These repairs were faster, with the ship returning to the air by the end of the month.

Ultimately, the prototype would have its fuselage lengthened, its engine moved about one foot forward, and its bomb bay lengthened. Additional cowl flaps were also installed and cuffs added to the propeller. While many of these changes were efforts to address directional stability problems, ultimately the vertical stabilizer had to be enlarged to correct the problem.

Flight testing of the XSB2C-1 came to a tragic end on 21 December 1941 when the prototype broke up in the air during a demonstration flight near the new Curtiss plant in Columbus, Ohio. Remarkably the test pilot "Red" Hulse was able to parachute to safety.

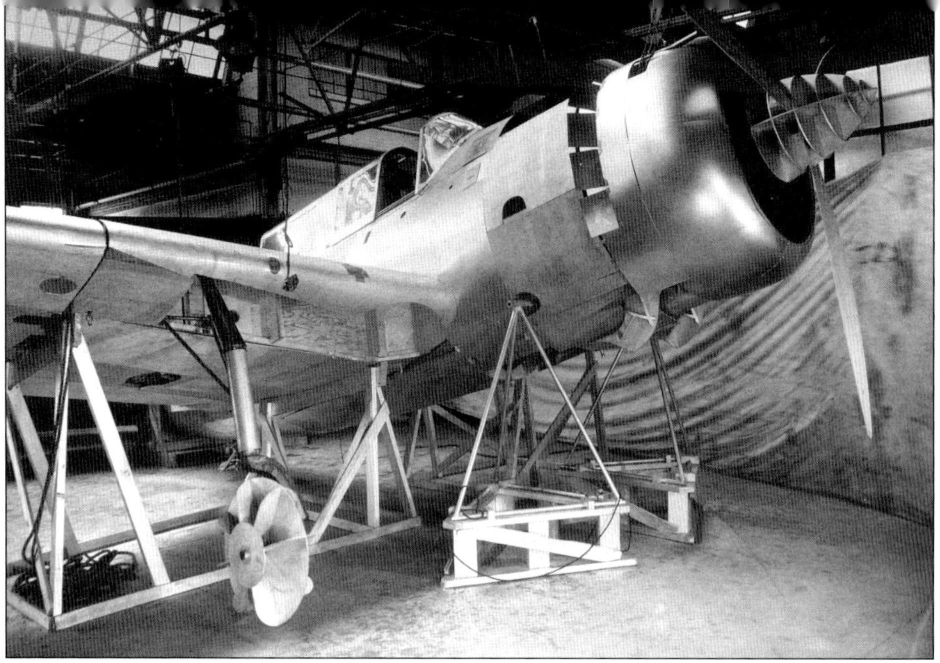

Less than a month after Curtiss received the contract to build the XSB2C-1, work on a full-scale mockup was well underway at the firm's Buffalo, New York, plant. The propeller spinner was in skeleton form at this point, and the cowl flaps are shown opened. (National Archives)

The mockup of the XSB2C-1 is viewed from the forward port side in a 16 June 1939 photograph. The initial design of the prototype included a turret in the aft cabin, and that turret is visible on the mockup. The outer port wing section had not yet been added. (National Archives)

The clear enclosure of the turret of the mockup of the XSB2C-1 is visible in this photograph taken on the port side of the fuselage. The machine gun, the gun cradle and mount, and other internal elements of the turret were not yet installed. To the left is the cockpit canopy; visible inside of it to the far left is a loop antenna for a direction-finding radio set. This radio used signals from two known transmitter sites to establish the aircraft's position using triangulation. The fuselage lacks some of the skin panels, allowing a view of the vertical formers and horizontal stringers of the frame. To the rear of the turret enclosure is the turtledeck, which was designed to be collapsible to give the turret machine gun a greater field of fire. Although the production SB2Cs would lack the turret, they would retain the collapsible turtledeck for the benefit of the gunner when operating the flexible machine gun. (National Archives)

A 16 June 1939 Curtiss factory photo shows the XSB2C-1 mockup from the starboard rear. The horizontal stabilizer was of a full-cantilever design. The starboard outer wing is in place. Split, non-perforated dive brakes/flaps were incorporated into the wings. (National Archives)

The flexible gun turret originally designed for the SB2C-1 is depicted in an 11 May 1942 factory photo. The Browning M2 .50-caliber machine gun is at 85-degrees elevation. The ammunition feed belt is on the left side of the machine gun, and to the right of the machine gun is an armor plate that afforded the gunner some degree of protection. Tests revealed problems with the gun turret, after further design revisions and the construction of additional mockups to try to fix the turret's issues, finally in the spring of 1942 the Bureau of Aeronautics (BuAer) allowed Curtiss to replace the turret with flexible twin-.30-caliber machine guns on a race mount. (National Archives)

Seen here resting on a hardstand, the Curtiss XSB2C-1 and early-production SB2C-1s had a three-bladed Curtiss Electric constant-speed propeller. A telescopic bomb sight is mounted through the windshield. The bulge aft of the cowl contained the exhaust.

The enclosed turret developed for the XSB2C-1 was not installed in the aircraft during flight tests. Fairings for the muzzles of two .50-caliber machine guns are on top of the cowl. In addition, there were two .50-caliber machine guns in the wings.

The XSB2C-1 performs a test flight on 26 May 1941. The prototype experienced stability problems from the beginning, and it suffered severe but repairable accidents on 9 February 1941 and a few weeks before this photo was taken, on 6 May 1941.

The XSB2C-1's stubby fuselage was owing to design constraints based on the BuAer requirement that two dive bombers with wings folded fit onto aircraft carrier elevators, for quicker delivery of the aircraft from the hangar deck to the flight deck.

XSB2C-1 Rebuilt

During a hard landing following a test flight on 9 February 1941, the XSB2C-1 suffered a broken fuselage forward of the empennage. The aircraft was repaired and resumed test flights on 6 May 1941. A series of modifications were made to the repaired prototype, including the addition of an air intake scoop at the chin of the cowling; the movement of the engine mount one foot forward, thus lengthening the fuselage; and several changes in the design of the empennage. By September 1941, the XSB2C-1 had a redesigned vertical fin and rudder, enlarged in an attempt to improve the plane's lateral stability.

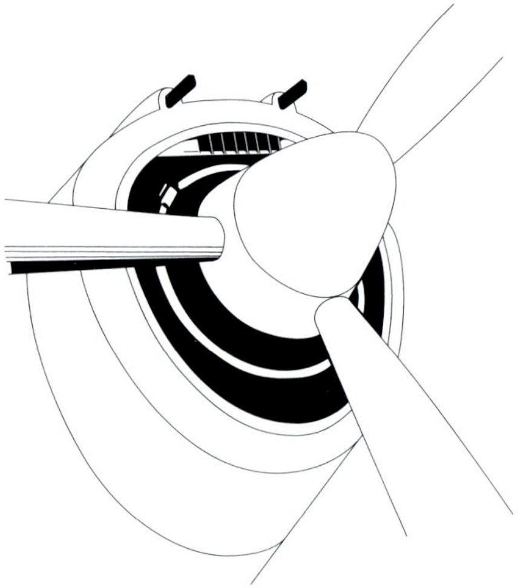

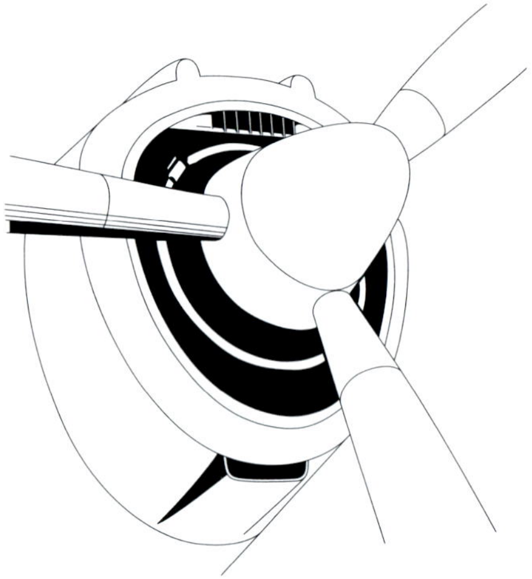

The two drawings above show the cowl of the XSB2C-1 before the 9 February 1941 crash and after the crash, when the chin air intake was added. The two drawings below show the original vertical fin and rudder and the one installed in September 1941.

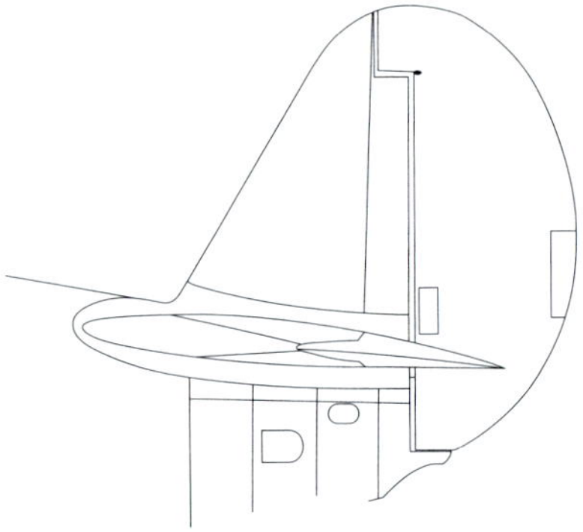

XSB2C-1

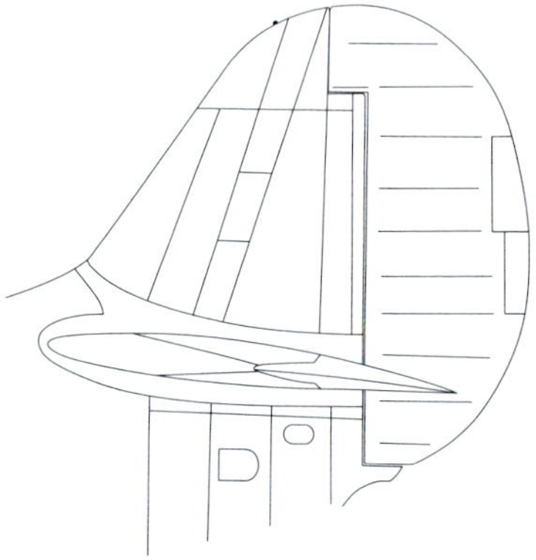

Rebuilt XSB2C-1

The nose of the XSB2C-1 is seen after the aircraft was repaired following the 9 February 1941 crash. Cuffs had been added to the propellers in an effort to improve cooling of the radial engine. The Curtiss Electric Propeller logo decals are visible on the type 714-1C2-12 propeller blades. In addition to the carburetor air intake at the top front of the cowl, an oil-cooler air intake was added to the chin, or lower front, of the cowl when the aircraft was repaired after the February 1941 crash.

With armed security guards standing by, crewmen roll out the XSB2C-1. The vertical fin and rudder are the final type, added in September 1941. During the 33rd test flight of the XSB2C-1, on 21 December 1941, the plane crashed, but pilot Barton Hulse bailed out.

SB2C-1

Completed in June 1942, the first production SB2C-1 is seen here on 2 July 1942. It was assigned BuNo 00001, making this aircraft the first one in the U.S. Navy's third series of Bureau Numbers. In all, 200 SB2C-1s were produced, and the number-one, SB2C-1, continued the test program begun by the ill-fated XSB2C-1. The SB2C-1 had a larger vertical tail than the repaired XSB2C-1 received in September 1941. In addition, on the SB2C-1, propeller cuffs were not installed, and the cowl machine guns were eliminated, and two .50-caliber machine guns were placed in each wing. A reflector gun sight replaced the pilot's telescopic sight. There is a radio direction-finder (RDF) loop antenna in the observer/gunner's cabin.

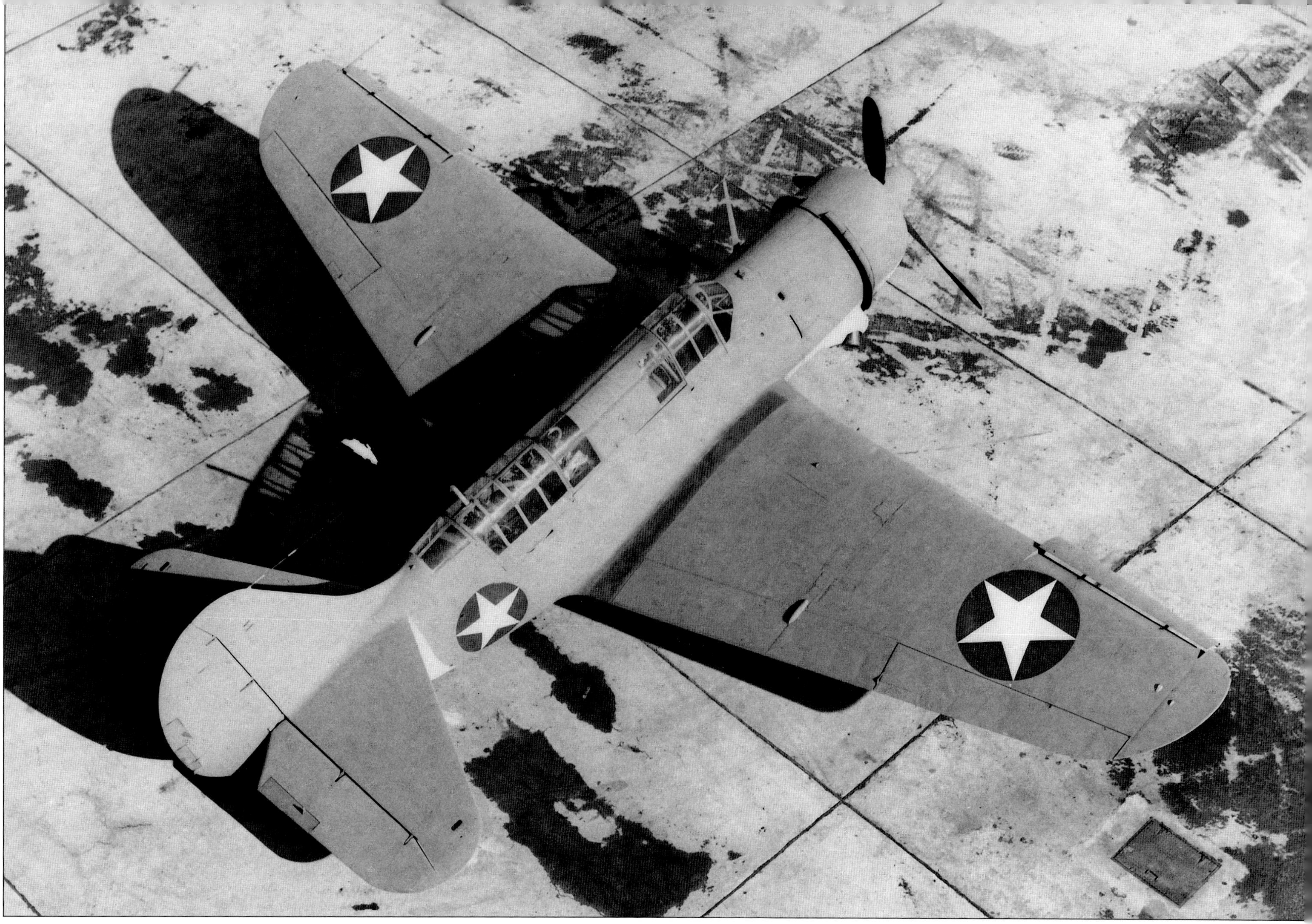

Several features common to early-production Helldivers are visible in this overhead view of an SB2C-1. The split flaps are not perforated; perforated flaps would not come until the SB2C-3. The pitot tube near the port wingtip is of the straight design found on the SB2C-1 (Mod 1); a bent pitot tube would replace the straight one beginning with the SB2C-1 (Mod 2). The trim tab on the starboard elevator was eliminated during SB2C-1C production. Aft of the sliding canopy of the cockpit on each side of the fuselage were several fixed windows; the narrow top window on each side would be deleted with the SB2C-3. The gunner had a single .50-caliber machine gun; its barrel is pointing up through a slot in the canopy.

An SB2C-1 wears a two-color camouflage paint scheme, most likely Nonspecular (NS) Blue-Gray over NS Light Gray, with NS Blue-Gray on the bottom of the outer wing panels. The national insignia is the design authorized from May 1942 to June 1943.

The oleo struts of the main landing gear on the SB2C-1 have a noticeable outward cant toward the top, giving the struts a bent appearance. The propeller tips have been painted in three colors, with a light-colored band between two darker ones.

SB2C-1 BuNo 00016 makes a flight over Ohio farmlands in January 1943. The exhaust fairing had undergone several modifications since the prototype, and the port exhaust stub protrudes through the fairing prominently. The RDF loop is visible in the aft cabin.

An SB2C-1 instrument panel yet to be mounted in a cockpit is depicted. The arrangement of the instruments would remain essentially the same until the SB2C-3. The assembly included an upper panel and a lower panel, mounted on shock-proof rubber bushings.

Between and slightly aft of the exhausts of this SB2C-1 is the open oil-cooler air outlet. This feature was not present on the prototype XSB2C-1. The number 00018, apparently the BuNo, is faintly visible on the vertical fin, roughly marked in a light color. (National Archives)

The SB2C-l shown in the preceding photo is viewed from a different angle. The number 7 is marked in white on the cowl and in black on the landing gear door. Protruding from the canopy over the aft cabin is the barrel of the single .50-caliber machine gun. (National Archives)

Experiments to mount torpedoes on Helldivers were conducted on SB2C-1 BuNo 00013 at NAS Quonset Point, Rhode Island, and BuNo 00149 at the Naval Proving Ground, Dahlgren, Virginia, resulting in an adapter kit that could be mounted in the bomb bay.

The number-five SB2C-1 comes in for a landing, arrestor hook lowered, on 9 July 1943. On the bottom of the wing is the marking 17-B-5, indicating the fifth aircraft of Bombing Squadron 17 (VB-17). The squadron's commander was Lt. Cdr. James "Moe" Vose.

SB2C-1 Mod II

The SB2C-1 was rushed into production, and as a result, it suffered many teething pains, not the least of which were stability problems. Modifications were undertaken to remedy these and other issues. At first, these modifications were performed in the field and were referred to as Mod I, but eventually a dedicated modification department was instituted at Curtiss's Columbus, Ohio, factory, where SB2C-1 Mod II aircraft were produced, beginning in April 1943. Mod II incorporated over 800 modifications, the most noticeable of which was the deletion of the RDF loop antenna in the rear cabin and the substitution of a bent pitot tube for the original straight pitot tube. A Yagi radar antenna is on the bottom of the starboard wing.

Helldivers finally became operational when VB-17 deployed on USS *Bunker Hill* in the summer of 1943. SB2C-1 Mod IIs are arrayed near a twin 5-inch/38-caliber gun mount. (Stan Piet collection)

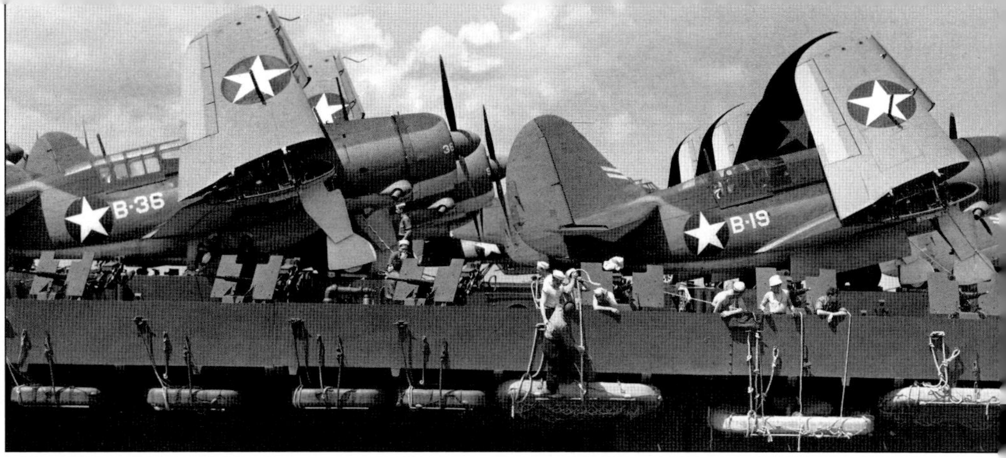

Curtiss SB2C-1 Mod IIs are lined up on the flight deck of a fleet aircraft carrier in mid-1943. Masts for Yagi radar arrays can be seen on the stars on the undersides of the wings.

A sailor installs a wing-locking bar, used to immobilize the folded wing, on an SB2C-1 Mod II. Yagi antennas are present under the wings of both Helldivers in the photo. (National Archives)

Deck crewmen and a lone pilot go about their business as the pilot in the cockpit of an SB2C-1 Mod II confers with a crewman during the early phase of Helldiver operations. (National Archives)

A SB2C-1 Mod II is on the aft elevator of the USS *Yorktown* (CV-10) during carrier qualifications in 1943. During the ship's shakedown cruise that May, there was a series of Helldiver crashes and accidents. Clearly, the Helldiver had yet to be perfected.

Aircraft of Carrier Air Group 5, including Helldivers of VB-5, are arrayed on USS *Yorktown* in 1943. The ship's captain, J. J. Clark, was disappointed in the Helldiver's performance and, following the shakedown cruise, replaced them with SBD-5s.

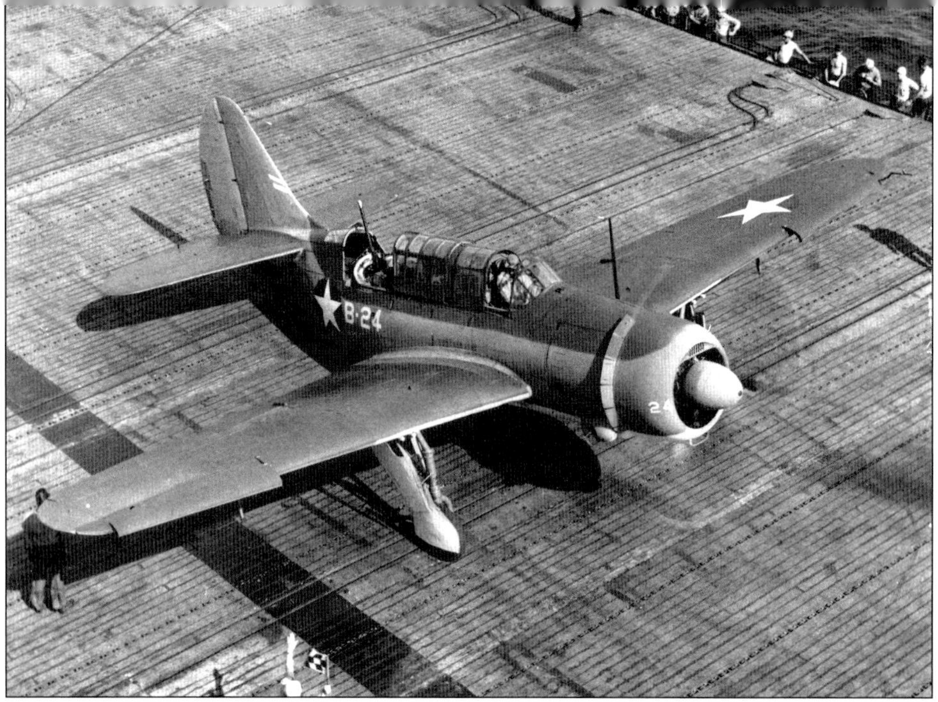

An SB2C-1 Mod II of VB-5 on USS *Yorktown* in June 1943 is painted in a three-color camouflage scheme authorized in January 1943, consisting of Sea Blue on the top surfaces, Intermediate Blue on the sides, and Insignia White on the bottom.

Fly One, the flight-deck officer in charge of launching aircraft, prepares to signal the pilot of an SB2C-1 to take off. To do so, he will drop to his right knee and point the checkered flag to the bow. Mounted to the front of the plane's windshield is a gun camera.

SB2C-1s of VB-6 warm their engines on the flight deck of USS *Hancock* (CV-19). Faintly visible on the vertical fins of the first two planes are the numbers 13 and 27. The turtledeck aft of the rear cabin on the first Helldiver has been lowered.

To the left, Fly One begins to dip to his knee to signal the pilot of an SB2C-1 of VB-8 to take off. Bombing Squadron 8 flew SB2C-1s while serving on the USS *Intrepid* (CV-11) during the carrier's shakedown operations from July 1943 to October 1944.

An SB2C-1 of VB-17 takes off from a carrier. A clear view is provided of the opening that was created aft of the rear cabin when the turtledeck was lowered, and the .50-caliber machine gun in the aft cabin, a characteristic of Mod III airplanes, is raised straight up.

During USS *Bunker Hill's* shakedown cruise in 1943, Bombing Squadron 17's SBC2-1s suffered a series of accidents, losing seven planes between May and October 1943. This Helldiver's entire empennage was torn off during an arrested landing on 8 July 1943.

A-25A

The U.S. Army Air Corps ordered their own version of the Helldiver. Given the designation A-25A, and popularly known as the Shrike, the Army version differed from the Navy variant in many details. Among them, but for the first 10, the Shrikes lacked folding wings. A-25A serial number 41-18774, shown here, exhibits a paint scheme of Olive Drab over Neutral Gray.

The U.S. Army Air Corps was not equipped with dive bombers when World War II erupted in Europe. Impressed with the combat record being chalked up by the Germans with their Stuka dive bombers, the Air Corps tendered an order in December 1940 for 100 SB2C-1As. Designated A-25A Shrike, the company planned to produce the Oive Drab bombers alongside the much more substantial Navy order at the Columbus plant. However, as the U.S. became involved in the war, and dive bombers continued to rack up a string of successes, the Air Corps orders were increased to a total of 3,000 Shrikes.

With the new total, the Air Material Command directed Curtiss-Wright to convert its massive St. Louis plant for Shrike production. Despite the outward similarity to the SB2C, there were a number of detail differences between the two aircraft, leading Curtiss to assign the Shrike its own Curtiss model number, S84, which in time was changed to 84B. The first A-25A lifted into the air on 20 September 1942. By March of 1943 the final aircraft of the 10-unit initial production order, which featured folding wings, had been completed. Subsequent aircraft were built with less-complicated fixed wings. Production of the Shrike continued to ramp up, but in June the Army decided to do away with all of its dive-bomber programs, preferring to concentrate its efforts in this vein on more versatile fighter-bombers. Massive contract terminations ensued and the 900 Shrikes that had been delivered were reclassified RA-25A – the "R" indicating Restricted to non-combat use. Ultimately, the U.S. Marine Corps acquired 410 of the Shrikes, redesignating them SB2C-1A Helldivers and installing Navy electronic gear in 1944. These aircraft, along with some from an aborted Australian order with the same designation, were used for training purposes.

The A-25A Shrike had the same straight pitot tube as the SB2C-1 Mod I. Wing slats are present, but these eventually were immobilized and later deleted. Radar was not installed in A-25As. The exhaust stubs on this example are longer than those on the SB2C-1. (National Archives)

The pilot of A-25A serial number 41-18774 poses in his cockpit. The exhaust stub on this aircraft is considerably shorter than the one in the preceding photo. A squiggle pattern of Medium Green paint has been applied to the leading edge of the vertical fin.

The long-type exhaust stub is visible on this A-25A. At the bottom of the fuselage just aft of the exhaust, the oil-cooler air outlet flaps are open. Details of the access panels and panel lines on the cowl and the forward part of the fuselage are also visible.

Workers at Curtiss St. Louis take a tour of the factory, getting a view the first of the A-25A Shrikes, serial number 41-18774. The aircraft is nearing completion and is resting on jack stands. This plant would produce 900 of these USAAF dive bombers in all.

Bomb-bay doors open, A-25A Shrike serial number 42-80194 makes a practice bombing run. The official designation of the production block to which this plane belonged was A-25A-30-CS. The "CS" in the designation referred to aircraft produced by Curtiss.

The pilot of A-25A-20-CS serial number 42-79800 prepares to enter the cockpit. The plane's nickname, "Torchy Tess," is written on the fuselage. The plane's tail number (the serial number minus its first digit) is painted in white on the bottom of the wing.

The layout of the instrument panel of the A-25A was unlike that of the Curtiss SB2C-1, shown earlier in this book. Rudder pedals are below the instrument panel, and in the foreground is the grip of the control stick. To the left are the armament switches.

The U.S. Army Air Forces did not use the A-25 Shrike in combat but turned it to other uses, such as towing targets and pilot training. These examples were used in training glider pilots. The Marines also used A-25s converted to SB2C-1As for trainers. (National Archives)

Curtiss A-25A-30-CS serial number 42-80171 sports a yellow band below the aft canopy and a yellow rudder. The Olive Drab paint is considerably weathered. This aircraft was surveyed and condemned on 1 May 1945. In the background is a B-25 Mitchell. (National Museum of the United States Air Force)

Resting at an airfield is A-25A-20-CS serial number 42-79804. A wavy pattern of paint, probably Medium Green, has been applied to the leading edge of the vertical fin. The national insignia has the red border authorized from June to September 1943. (National Archives)

This A-25A Shrike nicknamed "Gray Ghost" has nose art showing a skeleton wearing a hood. The bomb bay doors are slightly ajar, giving the belly a somewhat full appearance. The finish seems to be bare aluminum with a light-colored paint on the spinner and cowl. (National Archives)

The Royal Australian Air Force received 10 Shrikes, assigned RAAF numbers A69-1 through A69-10. In addition, further Shrikes bore RAAF numbers A69-11 to A69-81, including this one, A69-29, but these appear not to have been delivered to the RAAF.

Although the RAAF serial number is lost in the shadows on the fuselage, this appears to be another photograph of A69-29 making a test flight. The pilot and the gunner, hunkered down in behind the single .50-caliber machine gun, are both gazing at the cameraman.

The same Shrike in RAAF markings shown in the preceding photo, RAAF number A69-29, is seen from another perspective. General George C. Kenney, USAAF, commander of the Far East Air Force, effectively cancelled the RAAF's order for additional Shrikes.

One final shot shows RAAF serial number A69-29. Under the port wing, a bomb rack is visible. The tail landing gear was fixed, but without a fairing. As with other A-25A Shrikes, the tail wheel had a pneumatic tire, and no arrestor hook was installed.

SB2C-1C

The SB2C-1C replaced the SB2C-1 Mod III on the Curtiss assembly line. It incorporated the sum of the improvements that had been introduced in models I through III, along with a new armament package featuring 20mm cannons in lieu of the previous wing-mounted machine guns.

The changes inherent with the SB2C-1 Mod II still left a number of deficiencies, some so severe that the *Yorktown's* Captain "Jocko" Clark flatly refused to take the Helldiver into combat, preferring instead the venerable SBD-5. These problems led to the creation of the Mod III, a new model, which, among other modifications, had a fixed tail wheel with a streamlined fairing and substituted twin .30-caliber machine guns for the .50-caliber machine gun in the aft cabin.

Still, there was further room for improvement, and accordingly the SB2C-1C was introduced. The -1C incorporated the design improvements brought about gradually by its predecessors, and further augmented them with improvements of its own. Not the least of these was the upgrade of the fixed wing armament from twin .50-caliber machine guns in each wing to a single 20mm automatic cannon in each wing. Those wings now boasted hydraulic flaps as well as the strengthened folding mechanism found in the Mod III.

Curtiss built a total of 778 of the SB2C-1C, and they were assigned Bureau Numbers 00201-00370, 01008-01208, and 18192-18598. In addition to the substitution of two 20mm cannons for the four .50-caliber machine guns in the wings other external differences between the 1C and its predecessors included the addition of an air-flow deflector to each side of the gunner's canopy. The deflector was intended to reduce air turbulence in the gunner's compartment when the canopy was open. Also, the trim tab was eliminated from the right elevator, as was ultimately the balance tab from the rudder. Less visible, but important, was the addition of a hydraulic boost system for the flaps.

The SB2C-1C in the preceding photograph, BuNo 00245, is seen from the rear. The air-flow deflectors and tail landing gear fairing are not installed. Late in the SB2C-1 production run, the balance tab above the rudder trim tab was deleted.

This somewhat weather-worn Helldiver shown in a 5 June 1944 photo is an SBF-1, a version of the SB2C-1C produced under license by Fairchild of Canada. Fairchild manufactured 50 of these aircraft, which received Bureau Numbers 31636 to 31685.

Canadian Car & Foundry Co. also produced Helldivers under license. Although this example, an SBW-1, a version of the SB2C-1C, bears a U.S. insignia, it was part of a consignment of SBW-1s to be delivered to the British, and it bears a British serial.

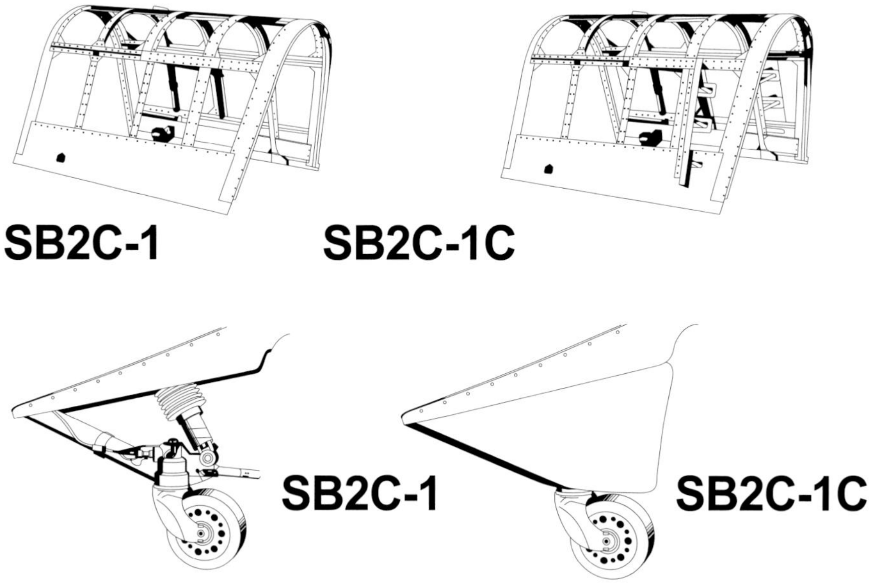

Several key design variations between the SB2C-1 and the SB2C-1C are illustrated. The top drawings show the addition of the retractable air-flow deflector to the gunner's canopy, and the bottom drawings indicate the addition of the fairing to the tail gear.

Two Helldivers attached to VB-1 on USS *Yorktown,* including an SB2C-1C, call sign "Jungle 70" in the foreground, fly a patrol near Guam in the summer of 1944. The diagonal stripe on the vertical fin was worn by Air Group 1 aircraft on the *Yorktown.*

An SB2C-1C assigned to VB-20 bears a three-color camouflage scheme and the code V35 on the fuselage. Yagi radar antennas are underneath the wings, and the air-flow deflectors are deployed in the open position on the gunner's open canopy.

An SB2C-1C assigned to VB-1 sets out for a strike on the Japanese airfield on Pagan Island in the Mariana Islands on 23 June 1944. Below are ships of Task Group 58.1. The gunner has trained his twin Browning .30-caliber machine guns slightly to the side.

Armorers pull torpedoes across a carrier deck toward an SB2C-1C. Although Helldivers had provisions for installing torpedo racks, this may have been a practice drill, since no documented cases of the delivery of torpedoes by Helldivers in combat are known.

A Helldiver observer/gunner captured this photo of SB2C-1Cs of VB-2 in June 1944. On the nearest aircraft, the fairing has been removed from the tail landing gear. The small white circle on the tail was the unit symbol for Air Group 2 during this period.

An SB2C-1C of VB-15 comes to a stop, with the arrestor hook still attached to the wire and the tail wheel in the air, on USS *Essex* (CV-9) in 1944. The pilot apparently failed to find a target on the preceding mission, as a bomb is still on the underwing rack.

As a result of a barrier-arrested landing on the USS *Essex* in June 1944, the nose forward of the cockpit of this SB2C-1C of VB-15 has sheared off. Although a fire-fighting crew is spraying fire-retardant on the smoldering wreck, the aircraft was a total loss.

During a raid on Chichi Jima on 6 July 1944, an SB2C-1C of VB-1 cruises high over an island group. A bomb, probably a 250-pound, is on the wing rack. Although the turtle deck is not lowered for firing the machine gun, the gunner is nonetheless on alert.

Aircraft number 18 of VB-2 on USS *Hornet,* a Curtiss SB2C-1C, bore a three-color camouflage scheme and risqué nose art in July 1944.

Helldivers of VB-8 warm their engines aboard USS *Bunker Hill* prior to a raid on Rota Island on 1 June 1944, and Fly One is about to flag the first plane in line to take off. A white horizontal bar above and below the tail number indicated aircraft of Air Group 8.

Handlers move an SB2C-1C of VB-8 to the spot on the deck of USS *Bunker Hill* where it will be secured in place. In the foreground is a tractor used for moving aircraft around quickly.

A sailor watches from atop of a twin 5-inch/38-caliber gun mount as a crew recovers an SB2C-1C that had an accident on the deck of USS *Bunker Hill.* The starboard landing gear seems to have collapsed, but otherwise the damage appears light. (National Archives)

An SB2C-1C of VB-14 has just nosed into one of the 5-inch gun mounts on USS *Wasp* on 31 August 1944. Fire-fighting crews are rushing into place. The pilot is still in the cockpit, and the damage to the plane appears not to have been catastrophic.

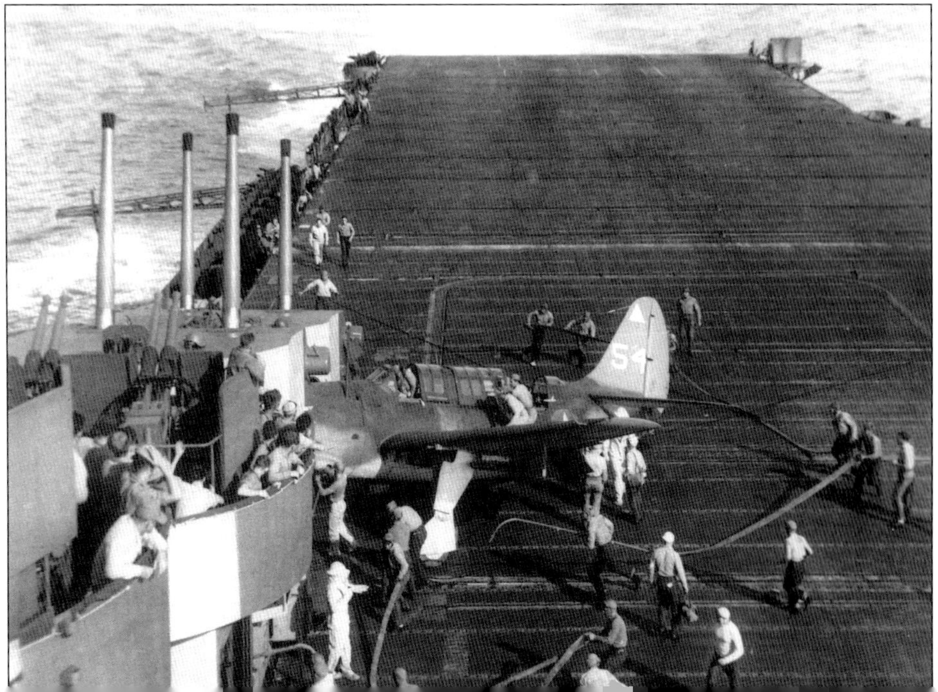

Well-weathered SB2C-1Cs are lined up at a base, wings folded. Judging from the taped-over carburetor-air intakes at the tops of the cowls, these Helldivers apparently were being held in storage. Different-colored propeller spinners signify varying units, with a mix of yellow, white, and blue spinners being visible. Details are available of the cowls, three-bladed Curtiss Electric propellers, main landing-gear doors, and exhausts and exhaust fairings. (Stan Piet collection)

XSB2C-2

In an effort to provide the Marine Corps with air support for amphibious operations, even if carriers were not available, the Navy launched a float plane initiative. Examples of all types of aircraft – scout, fighter, bomber were tested with floats toward this goal. Helldiver number 00005, the first Helldiver that was not hand-built, was selected to be modified with floats. After its delivery in conventional dive-bomber configuration it was ferried to the Anacostia Naval Air Station near Washington, DC, where it was fitted with a pair of Edo floats made for this purpose. When so modified, the aircraft was redesignated XSB2C-2.

Initial trials were conducted at Anacostia, where it was found that while satisfactory, the take off distance for the aircraft was more than had been hoped for. Nevertheless, testing continued with the float-equipped Helldiver going to Hampton Roads 9 March 1943 for rough water trials. These tests too went well, and the Navy ordered 294 examples of the float-equipped Helldiver. But the ongoing problem with the rear power turret with which the earlier aircraft were to be equipped hampered testing. By the time the decision had been made to utilize a twin .30 caliber flexible mount instead, the idea of a float-equipped aerial armada had waned. The XSB2C-2 project was cancelled 14 April 1944, by which time the U.S. had built, or was building, 110 of the 128 aircraft carriers the nation would arm itself with during the war.

Given the success of Japanese floatplanes with offensive capabilities early in World War II, the U.S. Navy ordered SB2C-1 BuNo 00005 fitted with Edo floats. It was designated the XSB2C-2 and was tested at Naval Air Station Anacostia, District of Columbia.

To compensate for the two large Edo floats, a ventral fin was installed below the rear part of the fuselage. The floats were of single-step design. In this photo dated 29 September 1942, beaching gear is attached to the floats. The bracing for the floats is visible.

The XSB2C-2 is viewed from the front. Tests of the floatplane indicated that it was generally a sound design but required long takeoff runs. Because of the evolving tactical situation as the war progressed, the Navy did not place this aircraft into production.

SB2C-3

The next model of the Helldiver, the SB2C-3, represented an effort to increase the power of the aircraft, replacing the Wright R-2600-8 engine, rated at 1,700 horsepower, with the Wright R-2600-20, rated at 1,900 horsepower. Along with the upgraded engine, the propeller was changed from the three-bladed Curtiss Electric to a four-bladed Curtiss Electric propeller with no spinner. In addition, Canadian Car & Foundry built 413 SBW-3s, and Fairchild of Canada produced 150 SBF-3s, both of which were based on the SB2C-3. This SB2C-3 was photographed at the Curtiss plant in St. Louis, Missouri, on 29 March 1944.

Another prominent change in the SB2C-3 was the modification of the sliding canopy over the gunner's compartment. At the rear of each side of the canopy frame was a streamlined fairing for a roller, which slid along rails on the outside of the compartment.

Fairchild of Canada manufactured under license 150 SBF-3 Helldivers to the same specifications as the Curtiss SB2C-3. This example, BuNo 31778, has the last three digits of the Bureau Number on the nose. The balance tab on the rudder had been discontinued.

Another in a series of photos of an SB2C-3 at Curtiss's St. Louis plant on 6 June 1944 shows to better advantage the prominent fairing for the roller at the rear of the aft sliding canopy. The aircraft had a decidedly pieced-together look, with a variety of finishes.

SB2C-3 BuNo 18602, the fifth production SB2C-3 to leave the assembly line, is viewed from aft. Although the outline of a trim tab on the right elevator, omitted during SB2C-1C production, is visible, close examination indicates that the tab had been immobilized. (National Museum of the United States Air Force)

The Wright R-2600-20 radial engine of an SB2C-3 is viewed close-up. The stencils on the Curtiss Electric propellers gave the drawing number, X714-8C2-12, the individual serial number of each propeller blade, and the high, low, and feathered propeller angles. At the top of the cowl is the air intake for the carburetor, and at the "chin" at the bottom of the cowl is the air intake for the oil cooler. (National Museum of the United States Air Force)

The cockpit of SB2C-3 BuNo 18602 is viewed from the starboard side. To the left are the throttle, supercharger, and fuel-mixture controls, below which are the bomb-release controls. The light colored box on the instrument panel above the grip of the control stick contains propeller controls. The autopilot controls are on the instrument panel directly to the front of the grip of the control stick. (National Museum of the United States Air Force)

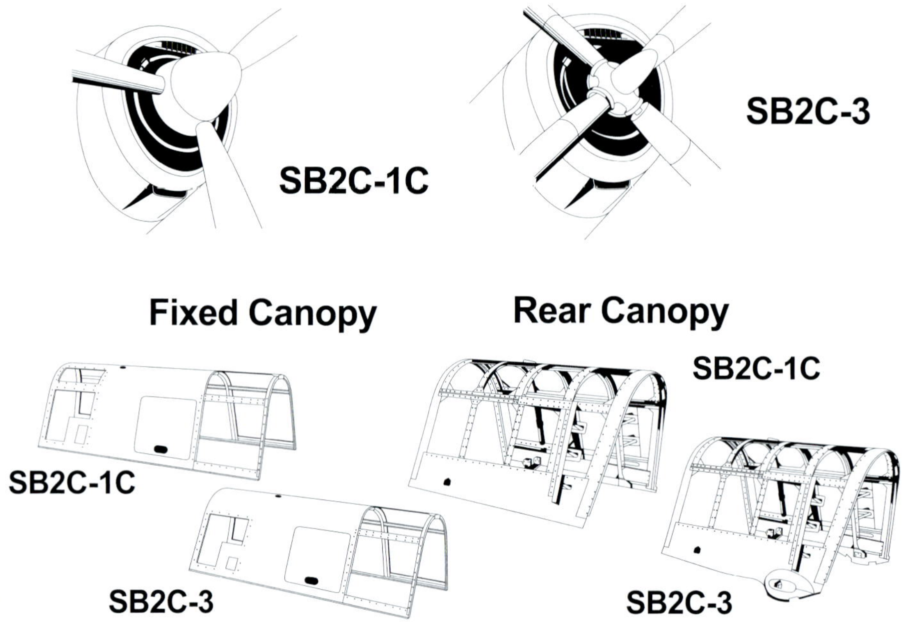

Canadian Car & Foundry manufactured under license 413 SBW-3 Helldivers, based on the SB2C-3. BuNo 21585, has the last three digits of the Bureau Number on the nose. The plane is painted overall in Glossy Sea Blue, a scheme authorized in 1944.

These diagrams trace changes from the SB2C-1C to the SB2C-3, including the switch to a four-bladed propeller without a spinner, the deletion of the small upper windows at the front of the fixed canopy, and the addition of the roller fairings to the aft canopy.

This SB2C-3 was assigned to the USS *Ticonderoga* in November 1944. On the vertical tail is the chevron symbol for the Ticonderoga's air group.

The Marine Corps flew a mix of different models of Helldivers in World War II. These were distributed over two dozen squadrons. Here, a formation of Marine Helldivers flies over Southern California around 1944. All but the nearest plane appear to be SB2C-3s.

Canadian Car & Foundry SBW-3 BuNo 21534 flies with a TBM-3 Avenger torpedo bomber near San Diego, California, on 26 February 1945. Helldivers and Avengers would work together on countless bombing missions from 1943 to the end of the war.

A Curtiss SB2C-3 Helldiver flies over mountainous terrain. The trim tab is no longer present on the starboard elevator, and the balance tab is likewise removed from the rudder. What appears to be a practice-bomb dispenser is on the starboard bomb rack.

The arrestor hook of an SB2C-3 has caught the arrestor wire, which will bring the aircraft to a fast and forceful halt. Two more arrestor wires in the foreground were available should the pilot miss the first one. The scene was USS *Tripoli* (CVE-64) in June 1944.

Three SB2C-3s fly in formation, the turtle decks lowered to give the gunners plenty of clearance for aiming their twin .30-caliber machine guns. All planes have Yagi antennas. Numbers are painted on the vertical fins: from front to rear they are 16, 18, and 19. The Helldivers bear the three-color camouflage scheme that the Navy adopted in January 1943, with Sea Blue on the upper surfaces, Intermediate Blue on the sides, and Insignia White on the bottom. The Intermediate Blue wrapped over the top front of the cowl and also appeared on the vertical tail and rudder.

Two USMC SB2C-3s fly over the Pacific toward the end of World War II. Conflicting sources identify these Helldivers as assigned to VMSB-151, based on Engebi Island in Eniwetok Atoll in the Marshall Islands, or to VMSB-343 "Gregory's Gorillas."

This photo and the one at right document the aftermath of an unsuccessful ground landing by an SB2C-3 in an overall Glossy Sea Blue paint scheme. The crash-landing probably was the result of a collapsed landing gear or failure of the gear to deploy.

An SB2C-3 painted in a late-war scheme of overall Glossy Sea Blue bears a stenciled number 6 on its vertical tail. The cowl flaps are open and both canopies are closed. The muzzles of the two 20mm cannons are clearly visible, protruding from the wings.

A view of the nose of the crashed SB2C-3 shows the destroyed propeller, with one blade completely missing. The blades had cuffs, which enhanced the cooling of the engine. In the background is a mobile crane, which will remove the Helldiver from the runway.

An SB2C-3 is ready for takeoff at NAS Wildwood, New Jersey, on 14 September 1944. Under the wing bomb rack is what appears to be a bomblet dispenser, which contained small bombs that provided an inexpensive means to practice bombing missions.

An SBW-3 Helldiver of VB-100 careened onto the catwalk of USS *Bataan* (CVL-29) during a landing on 25 January 1945. The arrestor hook is still holding on to the arrestor wire. With any luck, the plane would be repaired and returned to active service.

Escorted by a Grumman F6F Hellcat, an SB2C-3 assigned to VB-7, deployed with the USS *Hancock* (CV-19), flies off the coast of French Indochina (today Cambodia, Laos, and Vietnam) on 12 January 1945. On the vertical tail is the unit symbol of Air Group 7.

During the latter part of SB2C-3 production, the solid dive brakes/flaps were replaced by ones of a perforated design. At the same time, the upper dive brakes received a redesigned, crenelated trailing edge, which was designed to reduce tail buffeting.

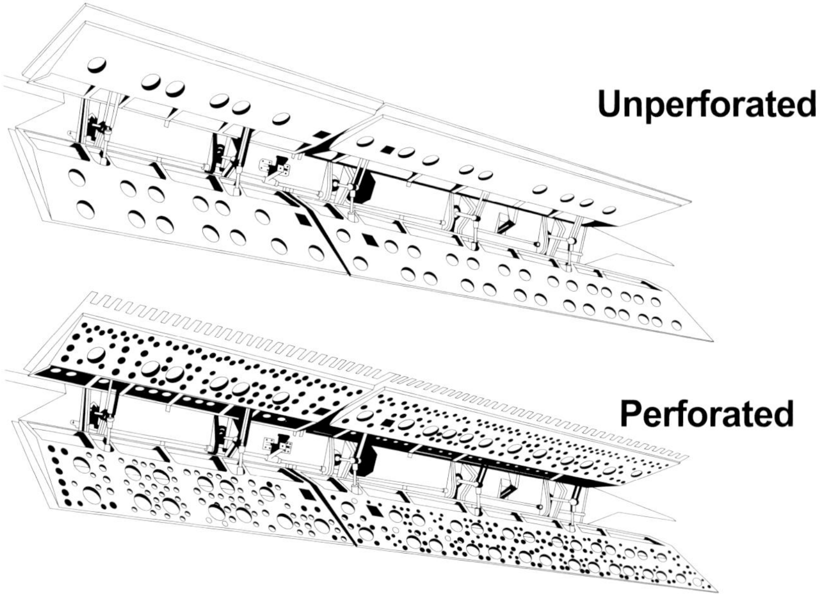

Three SB2C-3 Helldivers of VB-98 fly in formation near Ventura, California. This Navy bombing squadron never deployed to an aircraft carrier during World War II but rather functioned as a training squadron for Helldiver crews. The observer/gunner of the Helldiver to the right has lowered the turtledeck aft of his cabin, but the turtledecks on the other two aircraft are in the raised position.

An SB2C-3 of VB-7, based on USS *Hancock,* cruises off the coast of Taiwan ("Formosa") during air strikes on that island on 13 October 1944. The Hancock was part of Task Force 38, which in October 1944 conducted raids against Japanese air assets on Taiwan. Underneath the wing of this Helldiver is a pod containing two .30-caliber machine guns for strafing enemy targets when the opportunity was present.

As a landing signal officer (LSO) wielding a signal paddle turns to look, an SB2C-3 of VB-7 comes in for a landing on USS *Hancock* sometime in 1944. To the left is a windscreen of canvas lashed to a frame, which gave the LSO protection from the wind.

In the final months of World War II, an SB2C-3 of VB-6, flying off the USS *Hancock,* proceeds on a mission over land. The letter U was adopted as the tail symbol of aircraft of the Hancock's air group in the summer of 1945. A small drop tank is below the wing.

A deck crewman waves the pilot of an SB2C-3 of VB-7 into place on the flight deck of USS *Hancock* in 1944. The Yagi radar antenna loctated underneath the wing is turned outboard.

When a plane failed to catch an arrestor wire with its hook, barrier wires rigged across the flight deck would prevent the plane from crashing into planes parked forward on the deck. This SB2C-3 of VB-7 has hit barrier wires on USS *Hancock* and is flipping over.

A Curtiss SB2C-3 assigned to VB-7 on USS *Hancock* proceeds on a mission in 1944. A small bomb is shackled to the bomb rack on the underside of the wing. Visible in the foreground is the bent-type pitot tube on the other Helldiver from which an observer shot this photograph.

The SB2C-3E was similar to the SB2C-3, except the ASB radar with Yagi antennas was replaced by APS-4 radar with a sole radar-antenna pod on the starboard wing bomb rack. This SB2C-3E of VB-7 was photographed on the Hancock off Okinawa in October 1944.

In February 1945 an SB2C-3 of VB-9 based on USS *Lexington* (CV-16) flies high above ships of the U.S. fleet. This squadron participated on strikes against targets near Tokyo on February 16 and 17, including an attack on the Nakajima Ota aircraft plant.

An SB2C-3 of VB-15 prepares to alight on the deck of USS *Essex* (CV-9) sometime in 1944. On the flight deck in the foreground is a spacer in the shape of an inverted V, which held the arrestor wire above the deck, so the arrestor hook could snag it.

It is not clear why the nicely dressed woman is posing next to this Curtiss SB2C-4, BuNo 19951, but she adds a sense of scale and glamour to the dive bomber. This model of Helldiver was similar to the preceding one, the SB2C-3, except the -4 added a spinner to the four-bladed propeller, eliminated the side windows from the fixed cabin aft of the cockpit, and added four pairs of zero-length rocket launchers under each wing. Curtiss produced total of 2,045 SB2C-4s and SB2C-4Es, the former distinguished by the presence of Yagi antennas and the latter equipped with APS-4 radar. In addition, Canadian Car & Foundry made 270 SBW-4Es and 100 SBF-4Es, both of which were based on the SB2C-4E.

In a clear case of beauty and the beast, the young lady seen in the preceding photograph poses on the cowl of a Curtiss SB2C-4. Cuffs are present on the four propeller blades. Cowl flaps are open, as are the oil-cooler air-outlet flaps on the underside of the fuselage. Next to the woman's head is the cranked pitot tube, and a Yagi antenna is visible on the underside of the port wing.

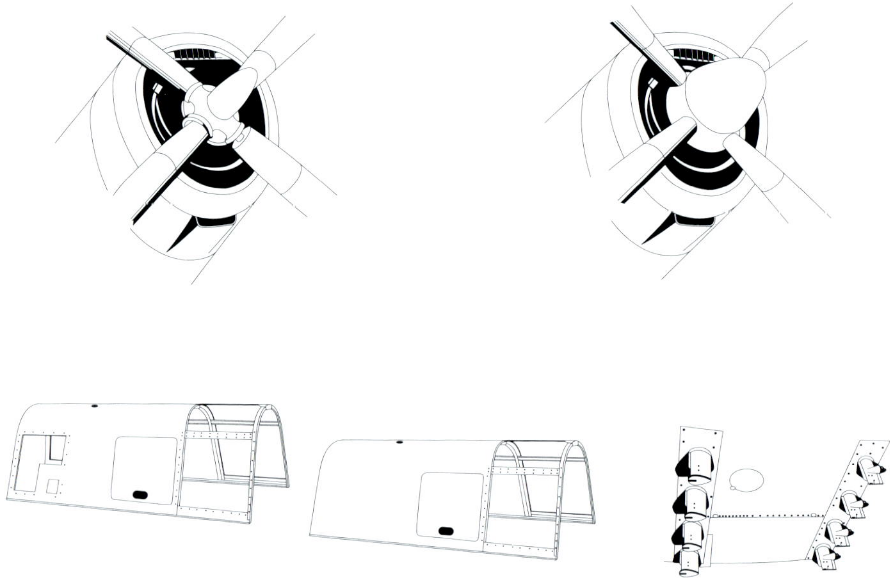

Three design changes differentiating the SB2C-4 from the -3 included the addition of a propeller spinner, the deletion of the side windows on the fixed conopy, and the factory installation of four sets of zero-length rocket launchers under each outer wing section.

The presence of a propeller spinner and the absence of a window on the fixed cowl identify this Helldiver as an SB2C-4. The plane has a tricolor paint scheme and no unit markings, and the turtledeck aft of the observer/gunner's compartment is lowered.

An SB2C-4, possibly the same one shown in the preceding photograph, is displayed in a wartime photo. The aircraft's clean condition, the presence of the farm fields below, and the fact that the observer/gunner is wearing an unmilitary long-sleeve white shirt suggest that this was a shakedown flight prior to delivery to the Navy.

The same SB2C-4 in the preceding photo is viewed from the port side. On late-production SB2C-4s the tail hook, hitherto housed in the rear of the fuselage until deployed, was relocated to a position outside of the fuselage. On this particualar aircraft, however, no tail hook appears to be present.

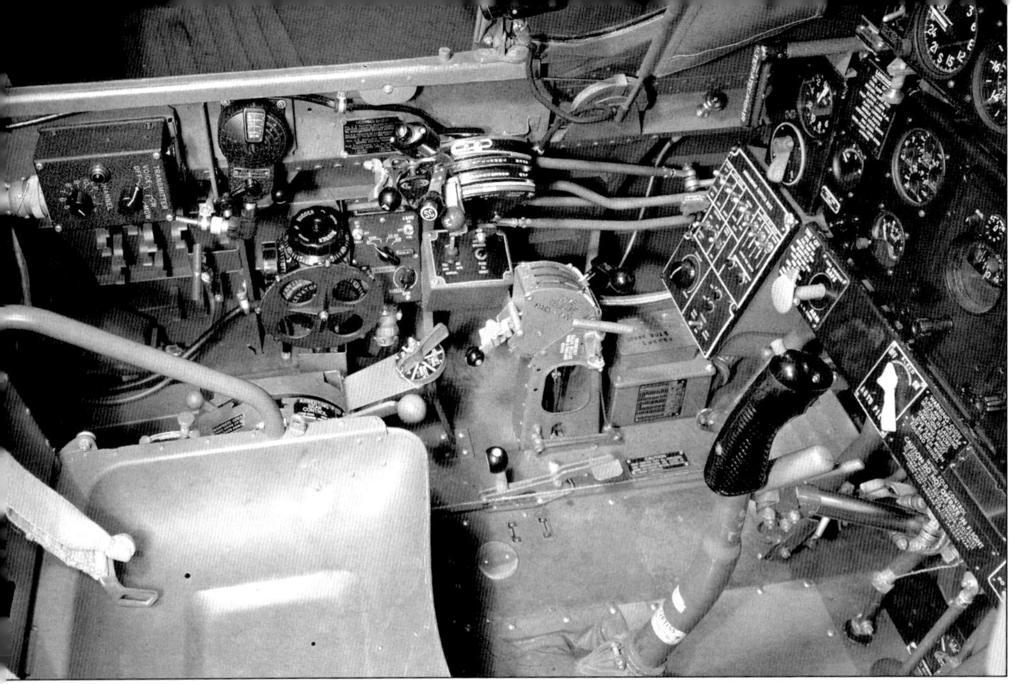

The port side of the cockpit of an SB2C-4 is displayed. To the lower left is part of the pilot's seat, showing the seat pan and a side brace. The small wheel to the side of the seat is the elevator trim tab control. (National Archives)

The handle with the ball grip next to the seat (on the starboard side of the SB2C-4 cockpit) is the handle of the hand pump that generates hydraulic pressure during a system failure. The hand crank to open and close the canopy is at top. (National Archives)

The pilot's instrument panel of the SB2C-4 remained a two-panel assembly, with some variation in the gauges from the preceding models. Although it is not shown in this photo, a sliding plotting board would have been installed between the upper and lower panels. (National Archives)

Radio equipment was in the forward part of the observer/gunner's of the SB2C-4. At the upper right is a shock-buffered rack for additional equipment. At the top is the bottom part of the life raft stowage tube. To the right is the hand crank for operating the canopy. (National Archives)

Atop the instrument panel of the SB2C-4 is an Illuminated Sight Mk. 8. A version of the RAF's Barr & Stroud GM2 gun sight, the Mk. 8 was a reflector-type sight that projected an aiming mark on the round glass screen, which was mounted at an angle of 45 degrees. (National Archives)

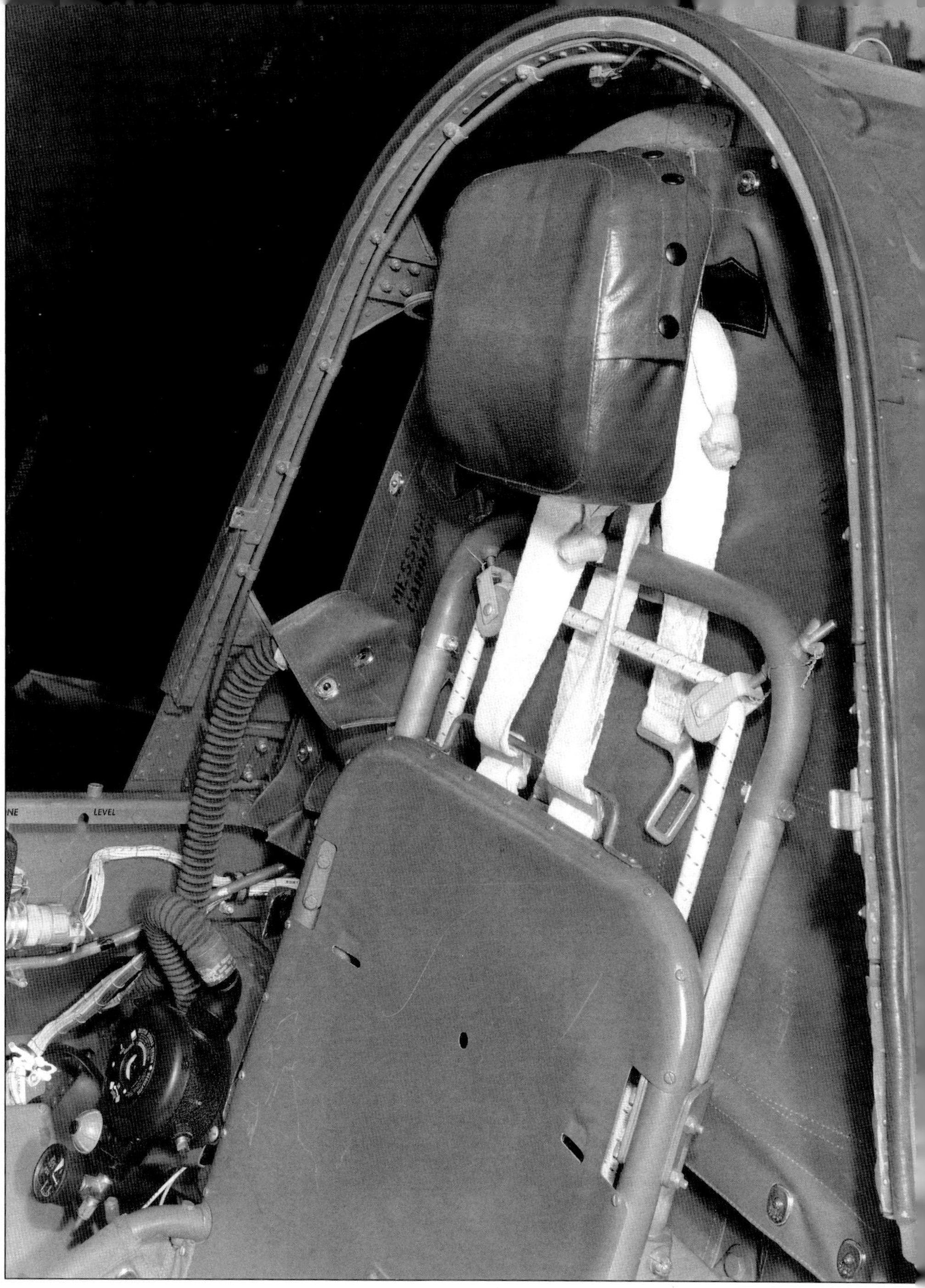

The upper rear of the cockpit of an SB2C-4 is shown, including the pilot's headrest, the upper part of his seat back and seat frame, safety harness straps, and oxygen hose. A removable canvas screen covers the opening into the fixed canopy behind the seat. (National Archives)

This photo of the observer/gunner's seat and machine gun ring mount, facing forward, was taken to document battle damage to SB2C-4 BuNo 20922, of a Lieutenant Roberts, Ulithi, 8 August 1944. A large tear in the seat has been marked with a circle. (National Archives)

The starboard side of the observer/gunner's cabin is viewed with the machine gun mount and gunner's seat removed. The starboard mount for the ring is the assembly at the upper center supported by a diagonal brace. Racks for ammunition boxes are to the right. (National Archives)

The turtleback of an SB2C-4 is shown in the raised position. The sliding canopy of the observer/gunner's cabin is open, and the rear of the cabin is to the lower left. The folding joints, internal frame members, and two top windows of the turtleback are visible. (National Archives)

The SB2C-4's turtleback is displayed in the lowered, or collapsed, configuration. A frame member along the longitudinal centerline of the turtleback separated the two top windows. As can be imagined, lowering the turtleback created some air turbulence. (National Archives)

An SB2C-4 prepares to take off on a training flight from an escort carrier. This was probably a carrier-qualifications flight, in which a pilot proved his ability to land on and take off from the flight deck of an aircraft carrier. Helldivers did not serve during combat operations on light carriers or escort carriers, but they would conduct training flights from them. The lower doors of the main landing gear have been removed as a safety precaution, reportedly so the doors would not be damaged should the wheels and landing gear engage the barrier upon landing. (Stan Piet collection)

Painted in a striking overall yellow scheme, this SB2C-4 or -4E served as a test plane at the Naval Air Modification Unit (NAMU), Johnsville, Pennsylvania, during World War II. As its name suggests, this unit developed modified fixes and improvements for USN aircraft, performing modification work on well over 1,300 planes. Among other things, the NAMU was instrumental in adapting radar to a number of naval aircraft, including the SB2C Helldiver. This SB2C-4 exhibits the externally mounted tail hook that was introduced during production of that model.

The all-yellow SB2C-4 or -4E assigned to the Naval Air Modification Unit at Johnsville, Pennyslvania, is viewed from the front, with the pilot entering the cockpit and a ground crewman standing by with a fire extinguisher should an engine fire develop upon starting the engine. The propellers have a semigloss black finish with yellow tips. This photograph gives provides a good view of the stance and angle of the main landing gear struts.

An SB2C-4E photographed at NAS Norfolk, Virginia, around the end of World War II appears upon first glance to be painted in an overall light color, but upon close inspection, it exhibits a faded two-color antisubmarine camouflage scheme.

This SB2C-4 has an unusual antenna installation, barely visible on the fuselage below the windshield. It is an antenna for an SCR-729 aircraft radar beacon and identification friend or foe (IFF) system. This radar/IFF system later was redesignated the AN/APN-2.

In another photo of the same SB2C-4 shown on the previous page, the antenna for the SCR-729 AN/APN-2 set is visible to the front of the wing. These antennas were used in pairs, and a similar one would have been mounted on the opposite side of the fuselage.

When completed at the factory, large, white numbers, representing the last three digits of the plane's Bureau Number, were applied to each side of the nose. The numbers were quite conspicuous and were painted over once the plane was delivered to the Navy.

An SB2C-4 devoid of visible markings except for the national insignia sits on a hardstand at an air base. The lack of side windows in the fixed canopy between the cockpit and the observer's cabin was a feature introduced with the SB2C-4 and is clearly visible here.

A factory-fresh SB2C-4, BuNo 82974, wears an overall finish of Glossy Sea Blue, and the last three digits of its Bureau Number are in white on the nose. The externally mounted arrestor hook, a feature of late-production SB2C-4s, is visible below the rudder.

The underside of the same SB2C-4 depicted in the preceding photograph is viewed during a shakedown flight near St. Louis on 5 March 1945. Present on the bottom of each wing is a bomb rack and four sets of zero-length rocket launchers. When high-velocity aircraft rockets (HVARs) were mounted on these launchers, the Helldiver acquired the capability to unleash on a target a barrage of eight 5-inch unguided missiles, similar in effect to a broadside from a naval destroyer. The small T-shaped antennas on the bottoms of the wings were radio altimeter antennas.

A Curtiss SB2C-4 is observed from above during a flight over the Pacific Ocean near Guam on 6 June 1945. Wear patterns from foot traffic give the top of the center wing assembly a distinctly different appearance than that of the outer, folding wing panels. The wingfold line on each wing presents a clear demarcation between heavy wear and light wear. The elevators were formed from doped fabric stretched over metal frames, and their appearance is darker than that of the neighboring metal-skinned horizontal stabilizers.

Failing to catch an arrestor wire, Ensign Lee Ray Jennings of VB-87 has stopped the engine of SB2C-4E BuNo 65017, the landing gear of which has just hit the barrier wires on USS *Ticonderoga* (CV-14). The date was 29 May 1945 en route to a raid on Kyushu.

In a photograph taken moments after the preceding one, Ensign Jennings's SB2C-4E has come to rest on its nose, and he and his observer/gunner are preparing to exit the aircraft. The starboard wingtip had struck the carrier's superstructure, spinning the plane around.

On the USS *Ticonderoga* on 6 June 1945, SB2C-4E BuNo 78282, piloted by an Ensign Brewitts of VB-87, crashed into the barrier, the cables of which can be seen snagged around the landing gear struts and doors. Brewitts is bracing himself for the crash.

Returning from a strike on Tokyo on 30 July 1945, SB2C-4E BuNo 20953, piloted by Lt. j.g. William Walter Timmis, has struck the barrier wires on USS *Ticonderoga.* Two weeks later, on 13 August, this plane would be lost during another Tokyo raid.

Curtiss SB2C-4 number 15 of VB-15 is shown in its colors and markings when it served on USS *Randolph* in February 1945.

An SB2C-4 of VB-3, assigned to USS *Yorktown* (CV-10), proceeds on a mission around April 1945. The white area with the diagonal edge on the tail signified the *Yorktown's* air group. The small white object below the windshield is the squadron's panther insignia.

It is 19 February 1945, D-day on Iwo Jima, and two Helldivers of VB-80 off USS *Hancock,* including the SB2C-4 in the foreground, fly high above the island. The nearer plane is painted overall Glossy Sea Blue, while the other has the three-color scheme.

Two Marine SB2C-4s, serving with VMSB-244 "Bombing Banshees" and operating from an air strip at Malabang, on Mindanao in the Philippines, proceed on a mission in support of U.S. Army troops operating near Davao, Mindanao, around May 1945.

An SB2C-4 from NAS Astoria, Oregon, cruises over a shoreline on 23 August 1945. The plane was painted overall Glossy Sea Blue, and the finish displays heavy weathering, especially on the wing tops.

Sometime in 1945 an SB2C-4 assigned to VB-9 takes off from the deck of the aircraft carrier USS *Yorktown*. In addition to the diagonal white air-group marking visible on the tail, the forward-pointing triangle on the starboard wing also signified the *Yorktown's* air group.

Aircraft warm their engines in preparation for a mission on USS *Yorktown* during operations in support of the Okinawa invasion in May 1945, including several SB2C-4s and -4Es of VB-9. To the left and in the background are several Avenger torpedo bombers.

Curtiss SB2C-4 number 104 of VB-3, USS *Lexington,* wore these colors and markings during the February 1945 Tokyo raids.

A Helldiver pilot poses together with his observer/gunner. The pilot is wearing a one-piece flight suit, over which are an inflatable life vest and parachute harness. The shoes are "boondocker" boots, and the pilot wears a flight helmet and goggles. The observer/gunner also wears a parachute and harness, flight helmet, and goggles, but his dress includes an enlisted man's work shirt and dungarees. The observer/gunner's right hand is resting on a small armored shield affixed to the twin .30-caliber machine-gun mount. (Stan Piet collection)

With the tail of Consolidated PBY-5A BuNo 48273 protruding into the top of the scene, a quantity of Curtiss SB2C-4s are lined up at a naval facility. A close examination of the photograph reveals that the Helldivers' folded wings are secured against wind damage by locking bars attached to the outer wing panels and the fuselage. The Helldivers are painted in mid-war three-color camouflage schemes. And at least some of them have large numbers painted on the sides of their noses in white. Chocks are fitted to their tires. (Stan Piet collection)

SB2C-4E

Similar to the SB2C-3E, SB2C-4s that were fitted with the APS-4 radar sets and the associated under-wing radar antenna pod were designated the SB2C-4E. A white radar pod is secured to the rack under the starboard wing of this SB2C-4E. Part of the original factory three-digit number on the nose, 064, is showing through the Glossy Sea Blue paint, although it is not clear if the paint has weathered off of the numbers or if an attempt was made to partially obscure them with hastily applied paint.

Canadian Car & Foundry produced 270 SBW-4Es, equivalent to the SB2C-4E. This one, BuNo V60166 assigned to Naval Air Training Command, New York, does not have the radar antenna pod mounted on the starboard wing rack, or the lower landing-gear doors.

The externally mounted arrestor hook introduced with the SB2C-4 and -4E is displayed. On preceding models of the Helldiver, the arrestor hook had been designed to retract into the fuselage, with just the hook at the end exposed when not lowered for landing. (American Aviation Historical Society)

SB2C-4Es of VB-87 fly a mission from USS *Ticonderoga* in 1945. A drop tank is on the rack under the wing of the nearest Helldiver. These auxiliary fuel tanks were often seen on Helldivers making long-range strikes into the Japanese home islands in 1945. (Naval History and Heritage Command)

Helldivers of VB-85 flying from USS *Shangri-La* (CV-38), such as this SB2C-4E photographed on 17 August 1945, conducted a series of devastating raids on the remnants of the Japanese battle fleet at Kure Harbor in the final weeks of the war. (Naval History and Heritage Command)

In a photograph taken on 18 May 1945, two SB2C-4Es flank an SB2C-5 in the middle. These planes, painted in overall Glossy Sea Blue, are thought to have been assigned to the Naval Air Ordnance Test Station (NAOTS), Fort Lauderdale, Florida.

An SB2C-4 or -4E of VB-94 flies over its carrier, USS *Lexington* (CV-16), around August 1945. Several months earlier, the system of geometric group symbols was discontinued and tail letters indicating the group, such as for H *Lexingtons's,* substituted.

Using tow bars attached to the main landing gear, a tractor pulls a Canadian Car & Foundry SBW-4E assigned to VB-74 on the flight deck of USS *Midway* (CVB-41) in the Atlantic around March 1946. A rudder lock is engaged around the vertical tail.

A photo dated 26 November 1944 shows an SB2C-4E of VB-87 in flight over the Caribbean. At the time, the squadron was assigned to USS *Randolph* (CV-16), which was undergoing its shakedown cruise off Trinidad. A small drop tank is below the wing.

The pilot of this SB2C-4E, Ens. F. Johnson, is about to make a crash landing on USS *Shangri-La* on 14 March 1946. The aircraft is in an extreme nose-up attitude as it approaches the fantail of the carrier, and it is clearly an accident about to happen.

In the next photo of the sequence, Ensign Johnson's SB2C-4E has descended slightly, with an increased angle of incidence, and is about to collide with the aft edge of the fantail of USS *Shangri-La.* The carrier's number, 38, is visible on the fantail.

Ensign Johnson's SB2C-4E has collided violently with the aft end of the flight deck. The force of the collision has snapped off the aft part of the fuselage and the empennage and has hurled the front end of the aircraft onto the deck. Things look grim for Johnson.

The remnants of Ensign Johnson's Helldiver are now skittering across the flight deck of USS *Shangri-La,* with the fixed canopy coming off and the fuselage broken forward of the wing. Almost miraculously, Ensign Johnson survived this crash uninjured.

Its radar antenna pod gleaming below the starboard wing, an SB2C-4E piloted by Ens. Vaughn E. Herbel, VB-87, has failed to catch an arrestor wire on USS *Randolph* on 5 December 1944. The plane is about to crash into the barrier wires at the lower right.

Ensign Herbel's SB2C-4E has just crashed into the barrier wires, and the forward impetus of the aircraft has caused the tail to rear up, as was often the case in barrier crashes. The propeller has stopped, and one of its blades has made contact with the flight deck.

Rather than flipping over, as happened sometimes in barrier crashes, the rear end of Ensign Herbel's SB2C-4E is falling back to the deck. The starboard wing has suffered some crumpling. Herbel and his observer/gunner, survived the barrier crash.

In July 1945, just weeks before Japan surrendered, SB2C-4Es of VB-87 defiantly streak past Mount Fuji, once a symbol of the might and perceived impenetrability of the Japanese home islands. Now, Helldivers were regularly striking targets in Japan.

Flying a Fairchild of Canada SBF-4E, the license-built equivalent of an SB2C-4E, Ens. R. L . Jacboson makes the 25,000th landing on USS *Solomons* (CVE-67) on 7 August 1945. The arrestor hook has caught the wire, and deck crewmen will secure the plane.

Curtiss SB2C-4E BuNo 65141 of VB-87 crashed into the ocean upon takeoff from USS *Ticonderoga* on 9 August 1945. The pilot, still wearing his parachute, is standing on the port wing while the observer/gunner has retrieved the inflatable life raft.

On 24 July 1945 Lt. j.g. E. L. Wheeler of VB-87 crashed this SB2C-4E into the barrier upon landing on USS *Ticonderoga*. In the crash, the top of the tail was torn off, and damage was incurred to the starboard elevator and stabilizer and the starboard aileron.

SB2C-4Es assigned to NAS Oakland, California, fly over the Bay Area in early January 1946. As Navy Reserve aircraft, these Helldivers have orange bands and aircraft numbers, but no national insignia, painted on the sides of their fuselages.

This row of newly minted Helldivers is thought to represent Canadian Car & Foundry SBW-4Es on an airfield in Minnesota *en route* to delivery to Curtiss's St. Louis plant in early 1945. Fluttering from the pitot tubes are ribbons for pulling off the tube covers.

More Canadian Car & Foundry SBW-4Es sit at an airfield in Minnesota in transit to St. Louis. The photo provides a good idea of the design of these numbers, with angular edges. The main-gear tires that are visible exhibit diamond tread patterns.

An SB2C-4E of an unknown unit in 1946 exhibits a two-color camouflage scheme used in antisubmarine operations in World War II.

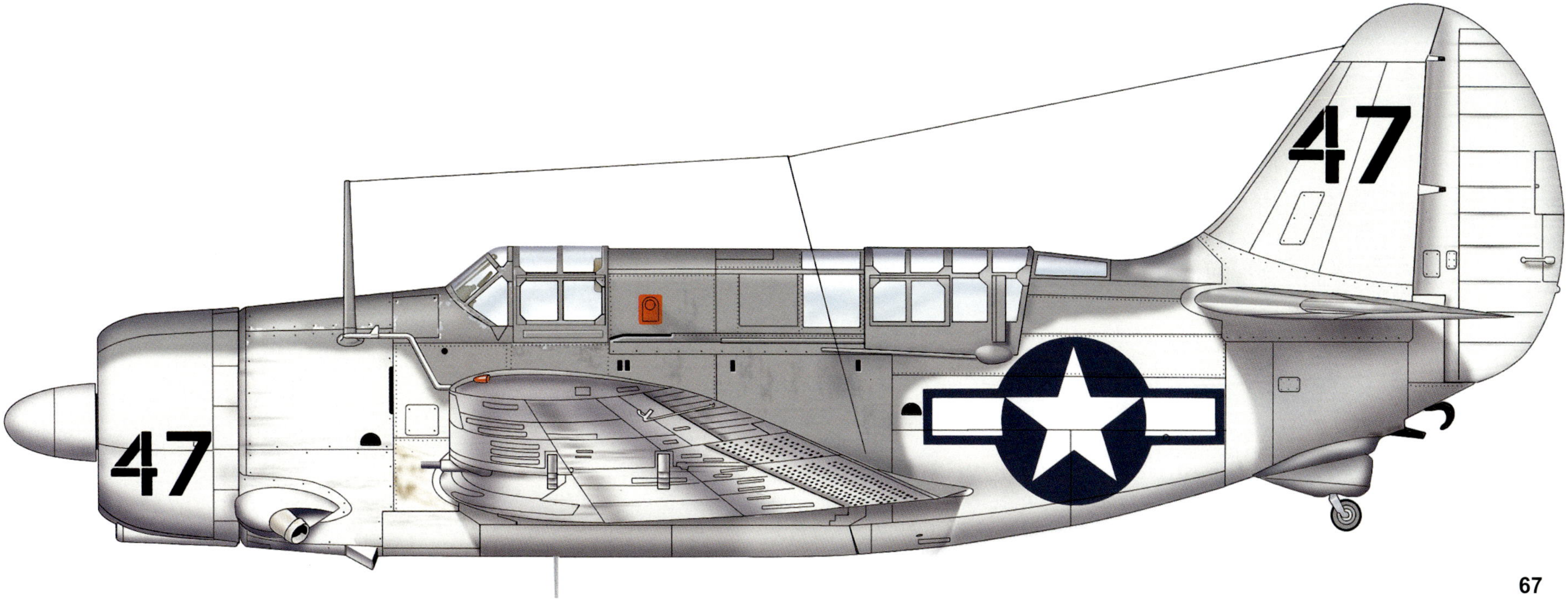

The SB2C-5 was the last production model of the Helldiver. The principal external feature that differentiated the -5 model from the SB2C-4 and preceding models was the redesigned pilot's sliding canopy, which lacked cross-framing. The -5 model introduced paddle-type blades to the propellers. Fuel capacity was increased by 35 gallons, and the bomb bay was enlarged and fitted with new doors with stepped front edges. Curtiss produced 970 SB2C-5s, and Canadian Car & Foundry produced 85 equivalents, the SBW-5. Seen here is an SB2C-5 assigned to NAS Glenview, Illinois, displaying an orange band denoting a Navy Reserve aircraft.

A new SB2C-5 undergoes a shakedown flight. The lack of cross-framing on the pilot's sliding canopy is obvious; this feature provided the pilot with better visibility to the sides. SB2C-5s retained the externally mounted tailhook, a featured introduced with the -4 type.

These diagrams illustrate the change from the presence of a propeller spinner on the SB2C-4 to the deletion of the spinner and addition of paddle-bladed propeller on the -5, and the redesigned pilot's canopy that eliminated the cross-frame members.

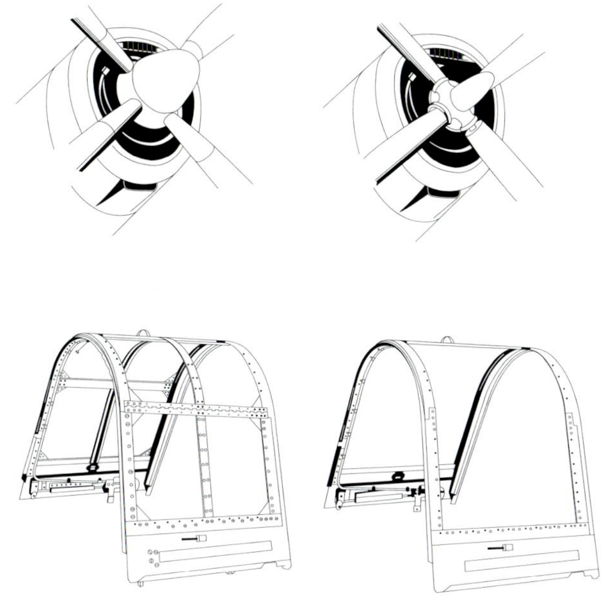

This lineup of aircraft at NAS North Island, San Diego, California, in 1946 includes, from bottom to top, an FJ-1 Fury, an SB2C-5, an F4U Corsair, and another SB2C-5. On each of the SB2C-5s, a white APS-4 radar pod is visible under the starboard wing, and on the nearer SB2C-5, orange bands identifying this as a U.S. Navy Reserve aircraft are present on the vertical tail and the wings. Both Helldivers have rocket-launcher rails instead of zero-length rocket launchers.

Crewmen of 5-inch/38-caliber gun mounts on USS *Philippine Sea* (CV-47) look on as an SB2C-5 of Attack Squadron 9A (VA-9A) lands in or around 1948. To the left, deck crewmen are running out preparatory to unfastening the arrestor wire from the hook.

The letter W underneath the wing on this Navy Reserve SB2C-5 indicates that it was assigned to NAS Willow Grove. An insignia that appears to incorporate a shield with wings to the sides is faintly visible on the fuselage to the front of the cockpit.

On this U.S. Navy Reserve SB2C-5 photographed around 1948, the U on the tail indicated that the plane was assigned to NAS St. Louis, Missouri, and the A stood for an attack aircraft. The Bureau Number, 82912, is aft of the national insignia on the fuselage.

An 11 April 1945 photograph shows a newly manufactured SB2C-5, still wearing its factory-applied numbers on the nose. Cuffs have been installed on the paddle-type propeller blades. Details of the mechanisms inside the wingfold joint are visible.

In 1948, a Curtiss SB2C-5 Helldiver serving with VA-9A "Tough Kitties" is ascending on the aft elevator to the flight deck of USS *Philippine Sea* while other SB2C-5s move into launch position. The aircraft bear the PS tail code of the *Philippine Sea's* air group. Attack Squadron 9A made a cruise to the Mediterranean on the *Philippine Sea* from 9 February to 26 June 1948. In August of that year, VA-9A was redesignated VA-94, and it was disestablished in November 1949. Several details of the scene in the photograph are worthy of notice. The SB2C-5 to the far right is equipped with a white APS-4 radar pod under the wing. The Helldiver on the elevator already has its engine running so it can quickly move into position once the elevator comes level with the flight deck. The interior of that plane's wing-fold joint is visible and it painted a dark green. Partially surrounding the aircraft elevator are stanchions with guard rails. These rails were in sections, and they were lowered flush with the deck when the elevator reached the level of the flight deck. Lying against the deck at several places are arrestor wires. Several of the Helldivers at the rear of the flight deck are starting to lower their wings, and one Helldiver still has its wings folded.

An SB2C-5 rests on a tarmac at an undisclosed airfield. It is painted in an unusual, three-color camouflage scheme, with, apparently, white on the undersides, a medium color on the sides of the fuselage and leading edges of the wings, and a dark top color.

The French bought a number of Curtiss SB2C-5s from the U.S. Navy in 1950 and used them in their war against insurgents in French Indochina (Vietnam). On the carrier *Arromanches* in the Gulf of Tonkin in late 1953 are some of those SB2C-5s.

This SB2C-5 served on USS *Lake Champlain* (CV-39) in 1945 and has that ship's geometric air group symbol on the tail.

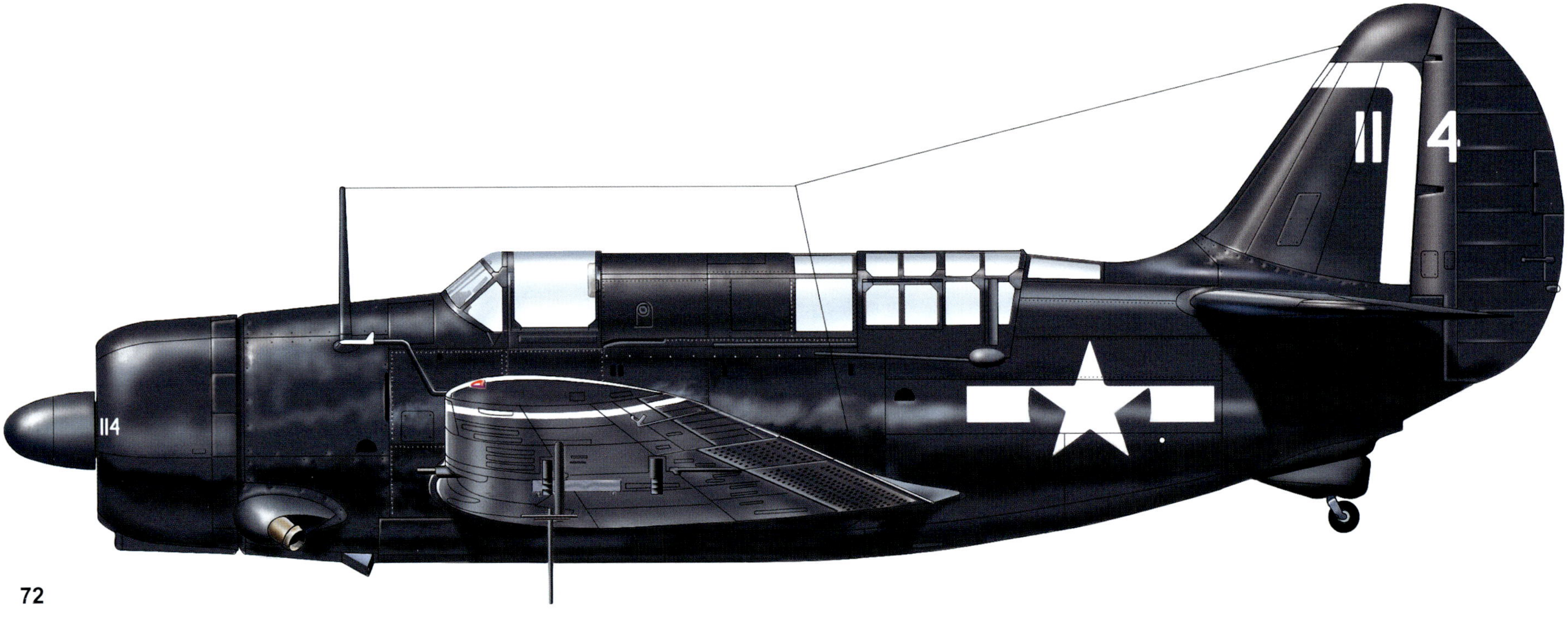

XSB2C-6

Even though the SB2C-3 through SB2C-5 had the Wright R-2600-20, rated at 1,900 horsepower, this engine was considered none too robust for the Helldivers. Therefore, plans were made to place the more powerful, 2,100-horsepower Wright R-2600-22 engine in the next model to have been produced after the SB2C-5: the SB2C-6. To that end, two prototype XSB2C-6s were made based on SB2C-1C airframes and were assigned Bureau Numbers 18620, seen here, and 18621. Production of Helldivers ceased at the end of World War II, and no SB2C-6s were manufactured.

Helldiver Production

Curtiss initially produced Helldivers at its Columbus, Ohio, plants. When the U.S. Army decided to acquire its own version of the Helldiver, the A-25A Shrike, Curtiss converted one of its existing plants, in St. Louis, Missouri, to produce Shrikes. In this photograph, Curtiss A-25As are undergoing construction on the St. Louis assembly line. The one in the foreground is nearing completion; various shades of zinc chromate primer are visible on the interior and exterior surfaces. To the far left is the very first A-25A, serial number 41-18774.

Curtiss's Columbus, Ohio, plant is busy makikng Helldivers. The aircraft were assembled on individual trolleys to facilitate moving them from station to station on the line. The two planes in the foreground lack canopies, while those farther down the line have them.

A Helldiver nears completion on the Curtiss Columbus plant assembly line on 4 February 1944. To spur on the workers to greater efforts and to boost morale, American flags and a production scoreboard for the month of February are prominently displayed.

Helldiver Data

Model	XSB2C-1	SB2C-1	SB2C-1C	SB2C-3	SB2C-4	SB2C-5
Length	36 ft, 8 in	35 ft, 5 in	36 ft, 8 in	36 ft, 8 in	36 ft, 8 in	36 ft, 8 in
Wingspan	49 ft, 8⅝ in	49 ft, 8⅝ in	49 ft, 8⅝ in	49 ft, 8⅝ in	49 ft, 8⅝ in	49 ft, 8⅝ in
Height, overall	16 ft, 11 in	16 ft, 11 in	16 ft, 11 in	16 ft, 11 in	16 ft, 11 in	16 ft, 11 in
Height, resting on gear	13 ft, 1½ in	13 ft, 1½ in	13 ft, 1½ in	13 ft, 1½ in	13 ft, 1½ in	13 ft, 1½ in
Engine	R-2600-8	R-2600-8	R-2600-8	R-2600-20	R-2600-20	R-2600-20
Horsepower	1,500	1,700	1,700	1,900	1,900	1,900
Maximum speed	325 m.p.h.	273 m.p.h.	273 m.p.h.	294 m.p.h.	296 m.p.h.	260 m.p.h.
Range	1,330 miles	1,375 miles	1,375 miles	1,200 miles	1,165 miles	1,845 miles
Ceiling	30,000 ft	21,200 ft	21,200 ft	25,000 ft	29,100 ft	26,400 ft
Weight, empty	7,122 lbs	10,114 lbs	10,114 lbs	10,114 lbs	10,547 lbs	10,580 lbs
Wing guns	2 .50 in cowl	4 x .50 cal	4 x .50 cal	2 x 20mm	2 x 20mm	2 x 20mm
Observer guns	1 x .50 cal	2 x .30 cal	2 x .30 cal	2 x .30 cal	2 x .30 cal	2 x .30 cal

The following sequence of photos was taken at Canadian Car & Foundry Ltd. plant, Fort William, Ontario, Canada, showing license-built SBW-1s, the equivalent of SB2C-1Cs, in production. Here, workmen apply themselves to nearly complete SBW-1s.

Wings folded, SBW-1s are taking shape on the assembly line. The ones on the right even have their Yagi radar antenna masts installed. At the bottom of the photo is a supply of three-bladed Curtiss Electric propellers, to the right of which is an engine assembly.

Canadian Car & Foundry technicians adjust Wright R-2600-8 engines destined for SBW-1 Helldivers. These powerplants were of radial design, with two rows of cylinders, seven cylinders per row. These engines could produce 1,700 horsepower at 2,600 RPM.

The fabric-covered elevators and rudders of these SBW-1s are a different color than the primed or painted aluminum-alloy skins. The trim and balance tabs had aluminum-alloy skins. Numerous access panels on the wings are open, their doors hanging by them.

A military officer by the cockpit joins Canadian Car & Foundry workers in inspecting an SBW-1 in production. On the wings are mats that helped protect the surfaces from scuffing. In the foreground, the interior of the engine-accessory compartment is visible.

A nearly complete SBW-1 undergoes inspections. The 20mm cannons will be mounted inside the wings later. The spinner has yet to be installed on the propeller, but the spinner's back plate is present. The bomb-bay doors are hanging open as well.

SB2C Bureau Numbers

XSB2C-1	1758
SB2C-1	00001 through 00004, 00006 through 00200
SB2C-1A	75218 through 75588, 76780 through 76818
SB2C-1C	00201-00370, 01008-01215, (01209-01215 cancelled, 18192 through 18307, 18309 through 18598
XSB2C-2	00005
SB2C-3	18622 through 19710
SB2C-3/3E	18599 through 18619
SB2C-4/4E	19711 through 21191, 64993 through 65285, 82858 through 83126
XSB2C-5	18308, 65286, 83127
SB2C-5	83128 through 83751, 89120 through 90019 (89466 through 90019 cancelled)
XSB2C-6	18620 through 18621
SBF-1	31636 through 31685
SBF-3	31686 through 31835
SBF-4E	31836 through 31935
SBW-1	21192 through 21231, 60010 through 60035
SBW-3	21633 through 21645
SBW-4E	21646 through 21741, 60036 through 60209
SBW-5	60210 through 60459 (60210, 60296-60459 cancelled), 21232 (21232 cancelled)

A-25A Serial Numbers

41-18774 through 41-18783	A-25A-1-CS
41-18784 through 41-18823	A-25A-5-CS
41-18824 through 41-18873	A-25A-10-CS
42-79663 through 42-79672	A-25A-10-CS
42-79673 through 42-79732	A-25A-15-CS
42-79733 through 42-79972	A-25A-20-CS
42-79933 through 42-80132	A-25A-25-CS
42-80133 through 42-80462	A-25A-30-CS

In Curtiss's St. Louis, Missouri, plant, A-25A Shrike dive bombers are under construction for the U.S. Army Air Forces. On the first Shrike in line, details of the Wright Cyclone R-2600 radial engine are visible as well as the port exhaust stub.

On the St. Louis assembly line, the tail of A-25A Shrike serial number 41-18775, the second A-25A to be produced. The next A-25A in line is supported by jack stands, and the panels have been removed from the engine-accessory compartment aft of the cowl.

These A-25As are apparently at the midway point in their construction, with the outer wing panels, among other assemblies, yet to be installed. On the nearest plane, the fixed cowling is not installed, giving a rare view of the rollover structure aft of the cockpit.

Workers at the Curtiss St. Louis plant, including a good share of "Rosie the Riveters," assemble inboard wing sections. The closest section is a starboard one, with the landing gear bay at the bottom. Each trolley holds two inboard wing assemblies.

Helldivers move along the assembly line at the newly-built Curtiss-operated plant in Columbus on 31 January 1943. While the Army variant was produced in St. Louis, and additional Navy-specification aircraft were built by Canadian Car and Foundry as well as Fairchild-Canada, by far the majority of the Helldivers were produced in Columbus.

The front of the aft cabin of an A-25A is viewed from the port side, showing the radio installations and, at the bottom, the machine-gun ring mount. This aircraft is equipped with the early type of sliding aft canopy installation that lacked the roller fairings and exterior rails.

Assigned to the 13th Fighter Squadron, U.S. Army Air Forces, this A-25A Shrike had an Olive Drab over Neutral Gray camouflage scheme.

A row of Curtiss SB2C-4 Helldivers faces Catalina flying boats to the right at a U.S. Navy facility toward the end of World War II. As a dive-bomber, the Helldiver compiled a mixed record. The aircraft was significantly faster than the dive-bomber it was intended to supersede, the Douglas SDB, and it took up less space on a carrier. However, the Helldiver suffered from an array of problems, from instability in flight to faulty design features, especially in the landing gear and hydraulic system. The Helldiver suffered from a high accident rate and was a tough bird to fly, earning it the nickname "the Beast." Unfortunately, very few examples of this formidable yet faulty dive-bomber survive.